Salah Guettala

Effect of infill walls on the behavior of BA structures

Salah Guettala

Effect of infill walls on the behavior of BA structures

ScienciaScripts

Imprint
Any brand names and product names mentioned in this book are subject to trademark, brand or patent protection and are trademarks or registered trademarks of their respective holders. The use of brand names, product names, common names, trade names, product descriptions etc. even without a particular marking in this work is in no way to be construed to mean that such names may be regarded as unrestricted in respect of trademark and brand protection legislation and could thus be used by anyone.

Cover image: www.ingimage.com

This book is a translation from the original published under ISBN 978-620-8-82564-5.

Publisher:
Sciencia Scripts
is a trademark of
Dodo Books Indian Ocean Ltd. and OmniScriptum S.R.L publishing group

120 High Road, East Finchley, London, N2 9ED, United Kingdom
Str. Armeneasca 28/1, office 1, Chisinau MD-2012, Republic of Moldova, Europe
Managing Directors: Ieva Konstantinova, Victoria Ursu
info@omniscriptum.com

Printed at: see last page
ISBN: 978-620-8-99061-9

CONTENTS

General introduction

Issues

Reinforced concrete structures with masonry infill walls are a common method of construction in seismic zones around the world, particularly in Mediterranean countries such as Algeria. Infill walls are generally not considered in the design process and treated as architectural (non-structural) components, although their presence has a significant impact on the seismic response for reinforced concrete portal frame structures. In order to study the seismic response of reinforced concrete portal structures with masonry infill walls, and to assess the contribution of infill walls to the linear and non-linear behaviour of these structures. This study is of crucial importance in highlighting the need for a thorough seismic assessment. This is not only an economic or social concern, but also a matter of civil safety. The results of this research should guide the introduction of standards and regulations aimed at strengthening the resistance of reinforced concrete structures with masonry infill walls against earthquakes. These guidelines must be integrated into urban planning and construction processes to ensure the protection of citizens and the durability of infrastructures in seismic zones. In addition, civil engineering professionals must be made aware of the importance of considering infill walls as structural elements from the initial design and construction phases. This will help to improve the resilience of buildings to earth tremors, thereby ensuring the safety of occupants.

In order to address this issue, this research proposes a new modelling approach, entitled "Macro-modelling approach incorporating a fibre-discretised plastic ball joint". This method aims to take into account the non-linear behaviour of structural elements by assigning resistance and deformation capacities to each component. In the ETABS programme, as suggested in this study, the portal frame element is represented as a linear element with linear elastic characteristics. However, the non-linear characteristics of each portal element are modelled as fibre-discrete plastic hinges. For the infill walls, the macro model used allows the significant use of one or more equivalent diagonal connecting rods due to their simplicity and efficiency from a computational point of view. However, the use of this model is hampered by the difficulty of defining the non-linear force-displacement characteristics of the equivalent diagonal links, due to the large number of variables linked to the mechanical and elastic properties of the materials. To overcome this problem, it was decided to use the fibre model for equivalent diagonal struts. In this approach, the force-displacement curve is substituted by a stress-strain curve for each fibre, enabling the non-linear behaviour of equivalent diagonal braces to be accurately described and the seismic response of infill walls to be simulated.

Aims of the book

In this research we will try to shed light on the effect of infill walls on the linear and non-linear behaviour of reinforced concrete structures using the Pushover technique. The main objective of this work is to evaluate the effects of the arrangement of infill walls in plan and/or in elevation on the seismic performance of reinforced concrete structures. Secondly, their effect on torsion in the structure must also be verified.

Map of the book

General introduction.

Chapter 1: This chapter presents a literature review on the seismic performance of infill walls, explaining the different modes of failure that can occur in masonry infill. In addition, the different techniques for their modelling are presented.

Chapter 2: This chapter presents the analysis methods, in particular the non-linear static analysis method (Pushover) used in this study.

Chapter 3: Based on experimental studies, we present a numerical investigation into the accuracy of using equivalent diagonal connecting rods (macro-model) with a discretised fibre section (fibre ball joint) for modelling reinforced concrete structures with infill walls.

Chapter 4: This part involves an analytical study of engineering models using the ETABS programme to assess the effects of the layout of infill walls in plan and/or elevation on the seismic performance of reinforced concrete structures and to verify their effect on torsion in the structure.

General conclusion.

Chapter 1: Bibliographical review

1.1 Introduction

Reinforced concrete structures with masonry infill walls are widespread and commonly used in construction throughout the world. The need for infill walls in structures arises naturally from the need to create separation between the interior space of buildings and the external environment. However, especially in the past, masonry infills were always considered as secondary elements and engineers did not include them in their models or calculations, simply because masonry infills had no definitions of their behaviour. In this chapter, a brief review of the literature is presented. This covers the seismic performance of infill walls by explaining the different modes of failure that can occur in masonry infill. This is followed by a review of numerical modelling methods for infill walls, with a focus on macro-modelling (i.e. the equivalent diagonal link) with two main approaches: stiffness methods and strength methods up to models that incorporate a fibre-discretised plastic hinge.

1.2. Seismic performance of walls

Infill walls are important elements in the seismic performance of buildings. They modify the lateral stiffness and resistance of reinforced concrete buildings and improve the seismic performance of buildings[1–3] . However, they are often neglected in numerical models and analyses, even though they are important for defining a performance level for the whole building. The performance of infill walls subjected to in-plane and out-of-plane loads is discussed in this section.

1.2.1. Behaviour of infill walls in plan

Several experimental and numerical studies[4–10] have been carried out to highlight the effect of the various geometric and mechanical parameters of the infill and portal frame on the response and modes of failure. Studies[11,12] have identified four main modes of failure of in-plane infill walls, as illustrated in Figure 1.1 :

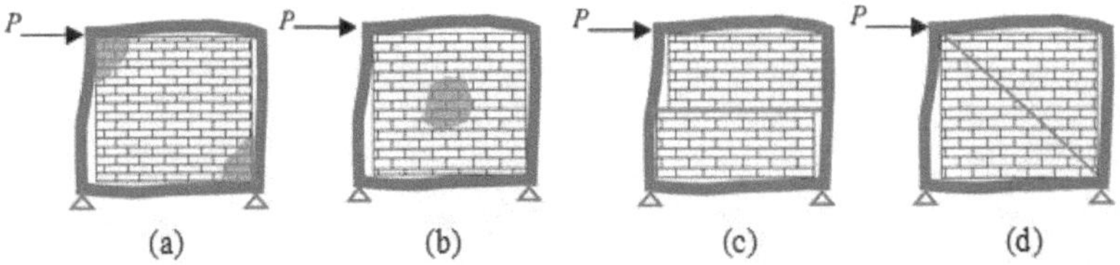

Figure 1.1 The modes of failure of infill walls in plan[12] : (a) compressive failure at the corners; (b) compressive failure at the centre of the infill; (c) shear failure at the centre of the infill along a mortar joint; (d) diagonal tensile failure.

In addition, the filled portal may fail due to the formation of plastic hinges in the beams and columns, or in the beam-post joints, as a result of the increased bending and shear forces caused by interaction with the masonry infills[5,13] . This mode is characteristic of weak porticos with solid infills. The formation of a gap between the portal frame and the infill cannot be considered a failure, but it does influence the performance of the system.

The occurrence of different failure modes depends on the geometry of the portico, the thickness of the infill and the material properties of the infill and portico. As indicated[12] , compressive failure occurs mainly in the corner regions and sliding shear failure at the mortar joints is the most significant. Diagonal cracking is not considered a mode of failure, because after cracking, the strength of the infill and filled portico continues to increase.

1.2.2. Behaviour of out-of-plane infill walls

Deflection in the out-of-plane direction of masonry infill walls is induced by inertia forces acting on the walls and by differential movement of the floors. Several studies[14–16] have investigated the behaviour of infills against out-of-plane loading. Figure 1.2a shows one of the failure modes where the entire wall tilts out of the portal frame due to weak gaps or connections between the wall and the surrounding portal frame. Other failure modes can be identified depending on the type of boundary conditions. Failure modes for a two-sided (horizontal direction, vertical direction) and four-sided supported infill are shown in Figure 1.2b-d.

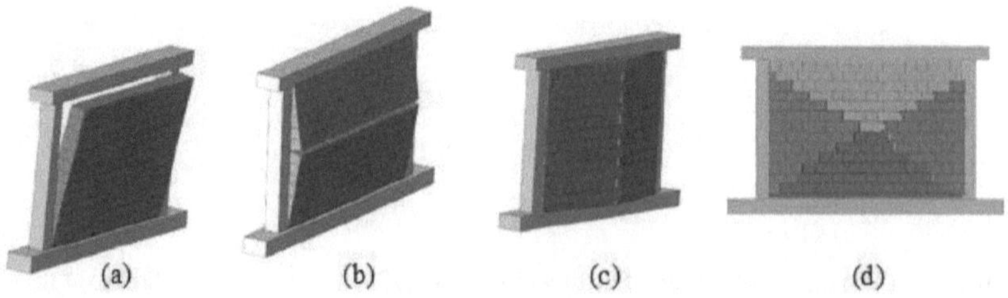

Figure 1.2 Failure modes of out-of-plane infill walls .[15]

1.2.3. Behaviour of infill walls under sequential loading

The significant influence of in-plane stresses on the out-of-plane strength of the infill was first observed by Paulay and Priestley[17] in 1992. In-plane behaviour is responsible for the reduction in out-of-plane strength due to the reduction in stiffness at the interacting surfaces between the infill and the portal frame. Again, gaps formed between the portal frame and the masonry wall are quite common and can be caused either by in-plane stresses or by shrinkage also responsible for influencing the out-of-plane load resistance.

The earthquake in L'Aquila (Italy) in 2009 demonstrated that most of the out-of-plane overturning collapses occurred in the lower storeys[18] . This can be explained by the fact that the overturning effects caused by the out-of-plane action are exacerbated by the in-plane deformation, causing detachment between the infill and the upper beam[19] . A typical X-shaped crack may form, weakening the infill and making the structure more vulnerable. Recently, several studies[20–23] have been carried out on the modelling of reinforced concrete porticos with masonry infill subjected to bi-directional loading, highlighting the importance of interaction effects between in-plane and out-of-plane loads. Other researchers[7,24–27] have also observed that in-plane damage can influence the collapse of infills under out-of-plane loading.

1.3. Experimental research

Experimental investigation is one of the most reliable tools that has been used to understand the behaviour of this type of structure in order to improve understanding of their modelling, analysis and design. Several experimental campaigns are listed in the literature and have been carried out over a period of six decades. These experimental campaigns have studied various aspects of the interaction between the reinforced concrete portal frame and the infill wall. However, this section focuses only on the experimental studies that analysed the influence of masonry infill walls on the in-plane behaviour of reinforced concrete portal frames. The experimental studies are presented in chronological order.

Polyakov[28] carried out the first experimental studies aimed at clarifying the interaction between portal frames and infill walls. Polyakov carried out experimental tests on steel portal frames filled with masonry. One of the main observations of this study is that the masonry infill wall functions as a bracing system for the portal frame by forming a compression rod. This observation motivated several researchers to use a compression rod as a model for the infill wall, which later became one of the main macro-modelling approaches for infills, as presented later in this chapter.

Later, Sachanski[29] carried out a series of monotonic static tests on full-scale models of reinforced concrete portal frames to calibrate its theoretical approach to assessing the contributions of infill walls to the combined stiffness and load distribution of structures. Sachanski used the experimental data to develop a theoretical method for analysing filled reinforced concrete frames. This method assumes that there is no separation between the portal frame and the wall, and that the wall has elastic, homogeneous and isotropic behaviour. However, these assumptions may not be entirely realistic.

Holmes[30] carried out several tests on filled steel portals of different sizes, subjected to tilting and shear loads. This study confirmed the theory of diagonal connecting rod formation. On the basis of his experimental work, Holmes proposed an equivalent connecting rod width "w" equal to a third of the diagonal length of the wall in order to calculate the maximum strength of the filled portal frame. A year later, Stafford Smith[31] also tested three steel portals with infills. This study was an extension of the equivalent connecting rod theory previously proposed by Polyakov[28] and led to an alternative method for calculating the equivalent connecting rod width based on the test results, which were used to measure the stiffness of the infill walls.

Fiorato et al.[32] carried out monotonic tests on twenty-seven reinforced concrete portal frames with masonry infills at a scale of 1:8. The specimens had different numbers of storeys, spans, reinforcement arrangements, gravity loads, concrete quality and wall openings. They concluded that reinforced concrete porticos with masonry infills have higher stiffness and strength, but lower ductility than bare porticos. Mainstone[33] and Mainstone et al.[34] also confirmed connecting rod formation in their experiments on large-scale concrete-coated steel frames filled with masonry walls. They proposed an empirical equation to calculate the effective crank width as a function of the relative stiffness of the frame and infill, and the diagonal length of the infill. To develop this empirical equation, they assumed that the effective connecting rod width is a function of the diagonal load on the infill at the first diagonal crack, the thickness and the compressive stress of the infill (measured from experiments). Later, on the basis of tests carried out on 1:3 porticos representing an eleven-storey reinforced concrete flat block from the 1970s, Klingner et al.[35] concluded that the presence of reinforced infill walls reduces the risk of incremental collapse, compared with that of a bare reinforced concrete portico.

In the 1980s, Zarnic et al.[36] carried out cyclical tests on twenty-eight specimens comprising bare and filled reinforced concrete porticos. The specimens had different scales, 1:2 and 1:3, infill materials (clay bricks and concrete blocks), types of openings (doors, windows but also solid infills) and also took account of unreinforced and reinforced walls. They found a significant increase in strength and stiffness, as well as a significant reduction in ductility, moving from bare porticos to filled porticos. They also reported a significant deterioration in strength after the filled systems reached their ultimate strength, and observed a severe reduction in the stiffness of the filled portal during cyclic loading. It is also important to mention that this study included the repair of nine of the specimens using different infill repair techniques (epoxy injection and a combination of epoxy injection with a reinforced cement lining) in order to investigate the effect of different strengthening approaches.

Later, Pires[37] tested seven models at a scale of 2:3, comprising single-storey, single-bay reinforced concrete porticos, six of which were filled with brick masonry walls and one bare portico. The bare portal model was used as a reference for analysing the results obtained from the filled models. The tests involved cyclic horizontal displacements imposed at the mid-axis of the beam and vertical forces applied at the top of the columns to represent the effect of gravitational loads. This study also included an analysis of the influence of certain parameters on the behaviour of the models, in particular those related to the construction procedures of the models, the reinforcement of the portals and the characteristics of the masonry. The study also addressed the influence of displacement on the failure mechanism of filled reinforced concrete portals.

Mehrabi[38] carried out twelve tests on half-storey and one-bay scale specimens, comprising two bare gantries and gantries with different filling conditions in terms of materials, loading procedures and reinforcement arrangement. After testing, the gantries were also repaired to analyse the effectiveness of the repair procedures. These test results were widely used by researchers to calibrate numerical models because of the comprehensive data available from the tests, including the data needed to develop refined finite element models and experimental explanations of failure mechanisms. These tests also confirmed the increased stiffness and strength of the filled portal frame compared with the bare portal frame.

Crisafulli[11] carried out a cyclic lateral loading test on two fully-filled reinforced concrete portal frames on a reduced scale of 3/4. The two specimens had the same size, materials and masonry type, but different reinforcement details. The use of tapered beam-post joints with diagonal reinforcement was discussed in one of these porticos. Crisafulli[11] concluded that the use of strong beam-beam joints improves the transfer of lateral force from the portal to the masonry infill. In addition, the test specimens showed significant crushing effects in their force-displacement responses.

The pseudo-dynamic algorithm presented in[39] was used to test a two-span, two-storey steel portal frame filled with unreinforced concrete block masonry to assess the seismic performance of the filled frame. The overall response and development of cracking patterns were similar to the results obtained from static tests carried out in previous studies. In addition, Buonopane et al.[40] carried out the same pseudo-dynamic test, but on half-scale reinforced concrete porticos with two openings on the second floor. It should be noted that, in this study, they observed the development of compression bracing for low force levels. However, for higher force levels, the diagonal pushrod contribution decreased and the stress distribution changed due to bed joint slippage at several locations in the wall. This change in the stress path in the infill implies the use of other strut configurations in the modelling, rather than simply a single diagonal strut, as will be presented in the following sections.

To study the influence of masonry on the strength and stiffness of reinforced concrete porticos, Al-Chaar et al.[7] tested, under monotonic static loading, a bare portico and four single-storey porticos filled at a scale of 1:2 with a different number of bays and masonry materials. They confirmed the significant increase in stiffness, residual strength and ultimate strength of the filled portal compared with the bare portal. In addition, they mentioned that increasing the number of spans implied an increase in the ultimate strength and stiffness of the filled portico, but the effect was non-linear.

Dynamic tests and monotonic static tests were carried out by Lee et al.[41] on non-ductile reinforced concrete porticos filled with masonry at a scale of 1:2 with two bays and three storeys. They also observed a significant increase in the stiffness, strength and inertia force (due to the added mass) of the filled portico compared with the bare portico. They also reported that the deformation capacity of the filled portico is almost the same as that of the bare portico. However, they concluded that the increase in inertial force due to the presence of the infill is less than the increase in strength, and they also observed a better structural response of the infilled portico in terms of peak strength and stiffness compared to that of a bare portico. They noted that, in terms of failure mechanisms, the corresponding bare portal experienced a soupel floor mechanism due to the formation of plastic hinges in the columns, and they also observed the shear failure of the columns in the lower level of the filled portal.

In order to observe the seismic behaviour of reinforced concrete porticos reinforced with precast concrete panels, Süsoy[42] tested eight specimens of reinforced concrete porticos at a scale of 1:3, of one storey and one span, with different panel types and connection designs. The specimens were subjected to cyclic lateral loading after being reinforced with precast concrete panels epoxy-bonded to the masonry infills. The effects of different panel shapes and connection techniques were studied. Süsoy[42] concluded that the reinforced infills failed due to excessive diagonal cracking in the panels, and the portal failed due to crushing, failure at the column bases or at the beam-post joints. After failure of the infill, the behaviour of the system became similar to that of a bare portal frame. Stronger infills provided higher lateral load capacity but hindered the action of the portal frame, limiting ductility.

Using a vibration table, Hashemi et al.[43] tested a 3:4 scale masonry-filled reinforced concrete portal frame as a model for a substructure of a five-storey prototype. They concluded that the presence of unreinforced masonry infills, in addition to their positive effect on stiffness and strength, also leads to an increase in the damping coefficient of 4-5-12%, depending on the excitation level, while shortening the natural period of the structure tested by 50%.

In order to assess the seismic performance of existing portal frames with and without infill walls, Pinto et al.[44] carried out a series of pseudo-dynamic tests on two full-scale models of a four-storey reinforced concrete portal frame representative of existing structures designed without specific seismic resistance features. The specimens were also retested after repair. They confirmed the high vulnerability of the bare portico and also reported that the filled portico showed a completely different behaviour to the bare portico. Although infills can protect the reinforced concrete structure, they promote the development of soft floor mechanisms and can lead to shear failure of the external columns in the joint regions. They also indicated that strengthening solutions based on 'K' bracing and dissipative devices, such as a shear link, can significantly improve floor behaviour and increase energy dissipation capacity.

In order to study the behaviour of ductile reinforced concrete porticos strengthened by the introduction of partial reinforced concrete infills under cyclic lateral loading, Anil et al.[45] constructed nine 1:3 specimens of one span and one storey and tested them under reverse cyclic loading. The aspect ratio of the infill wall (ratio of infill length to height) and its partial infill configuration were the main parameters of the study. This study indicated that partially filled reinforced concrete porticos also had significantly higher ultimate strength and higher initial stiffness than bare porticos. When the aspect ratio of the infill wall was increased, the lateral strength and stiffness increased significantly. In addition, they observed better strength performance for frames with connections between column, beam and infill. They reported that the fully-filled portal had an energy dissipation capacity seven times greater than the bare portal.

Kakaletsis et al.[46] and Kakaletsis[47] carried out a series of tests on specimens of single-storey, single-bay porticos at a scale of 1:3, subjected to horizontal cyclic loading to a displacement level of 4.0%. The specimens differed in terms of masonry materials (they used weak bricks and strong bricks), opening ratio to solid fill (from 0 to 100%) and also in terms of the location of the openings. They concluded that infills with openings can significantly improve the performance of reinforced concrete portals. In addition, specimens with strong fills showed better performance than those with weak fills.

Blackard et al.[48] carried out cyclic tests on six 2:3 scale specimens of masonry-filled non-ductile reinforced concrete porticos, replicating buildings from the 1920s in California. The main aspects of these specimens were the differences in the existence, location and configuration of the openings. In addition, the infill walls of two of the specimens were renovated with a specially designed cementitious composite material applied to the face of the walls. They observed that the rigidity and strength of the filled porticos were inversely proportional to the size of the opening.

More recently, Stylianidis[8] tested around forty specimens on a 1:3 scale. The experimental tests were divided into three campaigns. The first two campaigns focused on several aspects of reinforced concrete porticos with masonry infills, including mortar quality, the existence of a lintel beam, the contact condition between the reinforced concrete portico and the infill wall, material properties, the size of the masonry block and the aspect ratio of the portico. Ten damaged specimens from the first campaign were used in the third campaign, which analysed different reinforcement techniques. Of all the aspects addressed in this study, it was found that the aspect ratio of the portico has a minor effect on the behaviour of bare porticos but a major effect for filled specimens. In addition, Stylianidis reported that when the ratio of shear strength to compressive strength is high, exceeding a value of about 0.4, this can lead to brittle failure of reinforced concrete columns.

Stavridis et al.[49] also studied the dynamic behaviour of a 2:3 scale, two-bay, three-storey, masonry-filled, non-ductile reinforced concrete portal frame, representing a 1920s building in California, under scaled historical seismic acceleration records on a shaking table. This specimen was completely filled in one bay and had window openings in the other bays. They concluded that for seismic excitations less intense than a moderate design earthquake for the Los Angeles region, the structure behaved elastically. However, minor cracks appeared in the infill during a design seismic movement. As the intensity of the seismic movement increased, cracks in the masonry walls progressively propagated and caused the development of significant diagonal shear cracks in the reinforced concrete columns when the level of shaking, in terms of effective spectral intensity, exceeded the maximum earthquake considered by 43%. After this level of excitation, the damage to the structure was considerable but still repairable. As the intensity of the seismic movement increased further, the structure was severely damaged and large diagonal cracks developed in the columns of the first floor, causing a flexible floor mechanism.

In order to study the behaviour of reinforced concrete porticos with different types of masonry, namely standard masonry and interlocking masonry, Misir et al.[50] carried out quasi-static tests on three specimens at half scale; two reinforced concrete porticos fully filled with the type of masonry mentioned and one bare portico. They observed that interlocking brick infills maintain their integrity and stability up to very high levels of in-plane displacement, with less damage than conventional infills. As a result, they reported that interlocking masonry infills have the potential to reduce injuries and fatalities related to fall hazards during severe seismic shaking. However, they mentioned that the out-of-plane behaviour of locked masonry required further investigation.

To assess the effect of the existence of openings and their configuration on the structural aspects of portal frames with infills, Sigmund, et al.[51] tested six specimens of single-storey, single-span reinforced concrete portal frames. They tested a bare portal, a portal with a solid infill wall and the other specimens had infill walls with openings of different sizes, positions and types. To establish the behaviour of infill walls with openings, this study proposed correction factors for the behaviour of the bare portico, taking into account the type and position of the opening.

Mansouri et al.[52] also carried out a series of experimental tests to assess the influence of openings on the lateral behaviour of reinforced concrete porticos filled with low shear strength masonry. They tested six portico specimens at a scale of half, one storey and one span, under in-plane lateral loading. They mainly studied the influence of the shape, size and location of the openings. They concluded that the presence of openings alters the mode of failure, increases the level of damage and reduces the ductility, strength and stiffness of the filled portal frame. They also calculated the reduction in strength and energy dissipation capacity due to the existence of openings in the infill walls. Based on these tests, they proposed empirical equations to estimate the overall reductions in stiffness and strength of filled porticos due to the presence of openings, taking into account the size, shape and location of the openings.

Bergami et al.[53] tested three half-scale portal frames under cyclic lateral loads: one bare frame and two fully filled specimens. They reported that the contribution of the infill to the overall structural performance is significant due to the high stiffness and strength observed. They correlated the behaviour and damage pattern of the specimens tested with lateral displacement in four ranges: up to a displacement of 0.25%, there was no damage and the system behaved almost elastically; for displacements of 0.25% to 0.35%, they observed minor damage to the infill; for displacements of 0.35% to 0.6%, they observed constant lateral capacity and infill damage that was still repairable; above a displacement of 0.6%, the infill was severely damaged and its contribution should no longer be considered.

In the same context, Basha et al.[54] tested eleven half-scale specimens with full-scale and half-scale bricks. The tests were carried out in two campaigns. The first campaign included eight specimens, while the second campaign included three reinforced specimens. They reported that the specimens from the first campaign showed a significant increase in stiffness, strength and dissipated energy, which ranged from 7 to 10 times for stiffness, from 1.6 to 2.5 times for strength and from 1 to 2.3 times in terms of dissipated energy, compared with the behaviour of the corresponding bare portal frames. Based on their observations from the second campaign, they reported that the current provisions in the codes of practice do not appear to prevent shear failure in reinforced concrete columns of filled portal frames, even when using a weak fill with a strong frame. Therefore, they recommend improved design methods for reinforced concrete frames with masonry infills.

In order to simulate the failure of structural components due to abnormal loads or design faults, Shan et al.[55] tested two 1/3-scale, four-bay, two-storey reinforced concrete portal specimens without the first-storey central column, one a bare portal and the other with different infill wall configurations. Unlike previous tests, the portico specimens were pushed quasi-statically downwards at the top of the central column under controlled displacement. Even with this vertical loading, they reported that the filled portal frames had higher initial stiffness but lower ductility. They pointed out that in the case of large external loads or a fragile structure, the progressive collapse of the infill walls can lead to more damage to the structure. In addition, they observed the development of major cracks in the infill walls and cracks in the beams formed at an early stage for a very small deformation. Finally, they recommended reinforcing the beam regions near the corners of the opening to improve performance in the event of progressive collapse.

Using four full-scale specimens, one bare portal, one fully filled and two with partial fills, Zhai et al.[56] experimentally investigated the effect of fills and their different configurations on the overall behaviour of reinforced concrete structures. The specimens were tested under cyclic lateral loading. In addition to the high stiffness and strength recorded for the filled specimens, this study reported that the existence of the masonry infill wall caused cracks to develop in the columns at a lateral displacement 60-70% smaller than that of the bare portal frame. This implies that the infill accelerated crack formation in the reinforced concrete elements.

On the basis of the studies presented, it can be concluded that all the experimental studies have confirmed that the existence of infills, whatever their material properties, leads to a higher strength and stiffness than that of the corresponding bare portal, while reducing the available ductility, which, depending on the level of seismic intensity, can have positive or negative effects. In addition, most of these studies mentioned the formation of compression ties at low force levels. However, for higher force levels, the contribution of the diagonal link decreased and the stress distribution changed due to the occurrence of bed joint slippage at several locations in the wall[40] . Studies also report that filled porticos have a much higher energy dissipation capacity compared to bare porticos, as well as higher damping than bare porticos, while the ductility of the filled porticos is lower than that of the bare porticos[32,57] . However, Lee et al.[41] reported that filled porticos have almost the same deformation capacity as bare porticos. Based on these facts, the behaviour of masonry-filled reinforced concrete porticos can clearly be seen to involve different aspects than those analysed for bare reinforced concrete porticos. In addition, these tests also show that the variability in the behaviour of masonry-filled porticos, which depends, among other aspects, on the configuration and filling materials, is difficult to capture or generalise on the basis of limited experimental data.

1.4. Numerical modelling methods for infill walls

Due to the high cost of experimental testing, numerical studies have been carried out to simplify and determine the impact of masonry on the behaviour of reinforced concrete and steel frame portal frames at the design stage. Several techniques have been proposed for modelling infill walls. Researchers[14,58,59] have classified these methods into two categories: micro-modelling and macro-modelling. On the other hand[60,61] , have suggested three methods according to their complexity: micro-modelling, meso-modelling and macro-modelling.

These techniques differ in terms of complexity, accuracy and calculation time. The accuracy of the method varies in proportion to its complexity. Micro-modelling is considered to be the most complex method and therefore the most accurate. Macro modelling, on the other hand, is considered the simplest but less accurate. In this section, we present a brief bibliographical review of the numerical studies carried out for each of the three methods.

1.4.1. Micro-modelling

The method of modelling infill walls using detailed micro-modelling involves representing all the components of the infill walls, such as the bricks, mortar, brick-mortar interfaces and interfaces with the portal frame. This method was used by Riddington and Naom[62] in 1994 to predict the compressive strength of infill masonry. However, Lourenço and Rots[63] in 1997 simplified this method by replacing the mortar joints with elements of zero thickness and increasing the dimensions (width and height) of the adjacent bricks. This simplification is called simplified micro-modelling[60] . Figure 1.3 shows this transformation. Despite the simplification achieved by[63] , this method remains useful for research, as it allows to capture local phenomena and the failure modes of the various components. However, because of its complexity and high computation time, micro-modelling is not suitable for large structures.

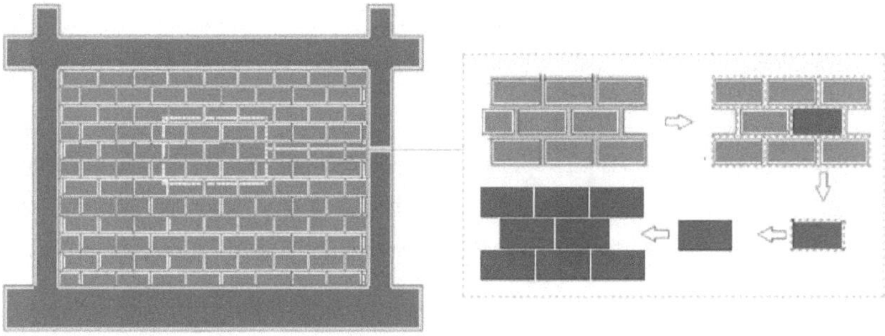

(a) (b)

Figure 1.3 Conversion from detailed micro-model to simplified model: (a) a portal with infill; (b) transformation by Lourenço and Rots .[63]

The modelling of masonry-filled porticos is a crucial aspect of structural analysis. Various researchers have proposed constitutive models aimed at accurately representing the complex behaviour of mortar joints and unit-mortar interfaces. These models, based on plasticity and integrated cracking, demonstrate a significant ability to predict the non-linear response of structures, offering promising prospects for improving design and seismic resilience. Below, a series of examples will be presented, reflecting the chronological evolution of approaches and advances in this field of research.

The constitutive model developed by Mehrabi et al.[5] aims to represent mortar joints and unit-mortar interfaces in finite element analyses. It takes into account compressive strain-hardening behaviour, shear expansion inversion and normal contraction of cementitious interfaces. The approximate results of the model compared to the experimental in terms of lateral strength and stiffness are observed, with a satisfactory match to the experimental failure mechanisms.

Oliveira and Lourenço[57] developed a constitutive model based on plasticity theory to represent the cyclic behaviour of interface elements. This model captures the characteristics of cyclic behaviour, stiffness degradation, energy dissipation and deformed patterns.

Al-Chaar et al.[64] used ADINA software to numerically study portal frames with masonry infill, showing the ability to predict the load-displacement curve and failure mechanisms. They recommend the use of appropriate material properties and structural geometry for optimal results using the finite element method.

Stavridis and Shing[65] have developed a micro-model to predict the behaviour of filled porticos by capturing the different modes of cracking. This model uses an integrated cracking approach to represent the brittle shear behaviour of the portico elements and the mixed cracking of the masonry mortar joint.

Koutromanos et al.[66] , following the approach of Stavridis[67] , combined an integrated and continuous cracking model with a cohesive cracking interface model. Comparison of the results shows an approximate prediction in terms of seismic response and failure mechanisms.

Sattar[68] has developed a 2-D micro-model to capture the cyclic behaviour of filled porticos, demonstrating the ability to predict important characteristics of masonry infill, such as cyclic shear slip and mortar joint cracking.

Mohyeddin et al.[69] proposed a 3-D micro-model to predict the response of porticos with in-plane and out-of-plane filling, using ANSYS. The results demonstrate the model's ability to capture the response of portals in both directions.

Al Louzi et al.[70] have developed a micro-model with ABAQUS to predict the response of reinforced concrete frames with infill. The model shows a satisfactory ability to predict the response of the filled portal frame under cyclic and monotonic loading.

Kubalski et al.[71] carried out a numerical study of reinforced concrete porticos filled with masonry, using LSDYNA and ABAQUS software. The numerical model was able to capture the overall behaviour of the system in both small and large deformations.

1.4.2. Meso-modelling

The previous method, which is micro-modelling, takes into account all the details of the filling in the model. However, this approach makes the modelling process more complex and difficult to carry out. Alternatively, the meso-modelling approach is less complex than micro-modelling. It is particularly useful for studying the behaviour of contacts between the portal frame and the filling. Meso-modelling makes it possible to simplify the heterogeneous composition of the masonry wall into a homogeneous composition. The constituent elements of the masonry infill are grouped together into a single equivalent element using the homogenisation technique. This technique is based on the selection of a small periodic unit cell that includes all the component types. The periodic unit cell is chosen so that its repetition generates the entire masonry wall[72] . Figure 1.4 illustrates the homogenisation procedure for a periodic masonry wall. Several homogenisation techniques have been developed and used by different researchers to determine the behaviour of infill walls .[73–80]

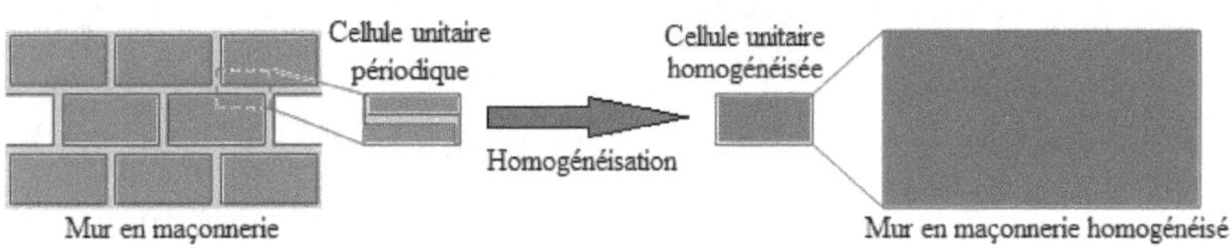

Figure 1.4 Homogenisation of masonry walls.

The study of behavioural models of masonry walls is a crucial step in the seismic design of structures. Through approaches such as homogenisation and meso-modelling, researchers aim to determine the equivalent properties and capture the failure mechanisms of these walls.

This summary presents a chronology of numerical and constitutive approaches, highlighting their contribution to understanding the complex behaviour of masonry walls.

Ma et al.[74] , through numerical simulation, used the concept of homogenisation on a representative volume element (RVE) to determine equivalent elastic properties and filling behaviour under various boundary conditions. They recommend an RVE as small as possible relative to the wall for increased accuracy, although the method may not always be applicable to various strains and stresses.

Milani et al.[75,76] presented a simple micromechanical model for walls subjected to horizontal in-plane loads, combining limit analysis and homogenisation. This model assumes that the masonry is subjected to in-plane stresses that develop a polynomial stress field in 2-D, showing an ability to represent the homogeneous failure of walls.

Yu et al.[81] have developed a numerical method based on the basic cell to determine the equivalent mechanical properties and failure characteristics of a masonry brick, showing an ability to capture the overall behaviour of the infill wall under different loadings, although the prediction of local damage is limited.

Sacco[82] proposed a non-linear constitutive model based on the mortar damage process and a non-linear homogenisation procedure to study the in-plane behaviour of masonry walls, suggesting a possible extension to include non-linear effects.

Milani and Lourenço[77] have applied meso-modelling to a large-scale masonry structure, using the unit cell concept to determine equivalent properties, although the model gives approximate results compared with other software using an elastoplastic model.

Quinteros et al.[72] used the homogenisation technique to solve masonry problems, adopting a linear elastic constitutive model and a one-sided damage model to simulate behaviour under cyclic loading, with the possibility of extension to include non-linear effects.

Houda et al.[83] carried out a numerical study on masonry walls with hollow concrete blocks, observing compression failures in the blocks and shear failures in the joint interfaces.

Finally, in the case of non-periodic masonry walls, Tiberti and Milani[79,80] applied the homogenisation technique with the test window concept to study the out-of-plane collapse behaviour of an ancient European structure with multi-leaf masonry walls.

1.4.3. Macro-modelling

Despite the simplifications proposed above, the complexity and calculation time required make both methods unsuitable for everyday use in the engineering field . For this reason, macro modelling of infill walls is a simpler and more practical method for taking into account the contributions of walls in large-scale structures. This approach involves simplifying the contribution of walls in terms of their behaviour. Although it does not capture the internal effects of infill walls, it is more suitable for studying overall behaviour. Macro modelling of infill walls is based on the representation of the walls by one or more diagonal rods in the load direction, as shown in Figure 1.5.

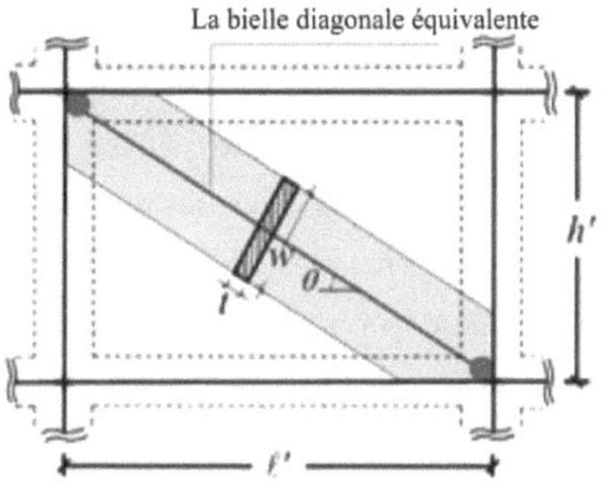

Figure 1.5 The equivalent diagonal connecting rod in the macro-modelling approach

One of the first macro-modelling approaches used to take into account the contribution of infills to the seismic response of reinforced concrete buildings was proposed by Polyakov[84] and implemented by Holmes[30] and Stafford Smith[85] in the 1960s. It consists of adding two diagonal rods opposite each other to the reinforced concrete portal frame (Figure 1.6a) and has been widely adopted by engineers and researchers, as shown by its inclusion in seismic codes[86] . However, as Crisafulli[11] points out, although the overall behaviour of reinforced concrete frames with masonry infill is fairly well predicted, single-rod models are unable to take account of the local interaction between the frame and the infill, which leads to inaccuracy in the distribution of shear and moment demands in the reinforced concrete frame.

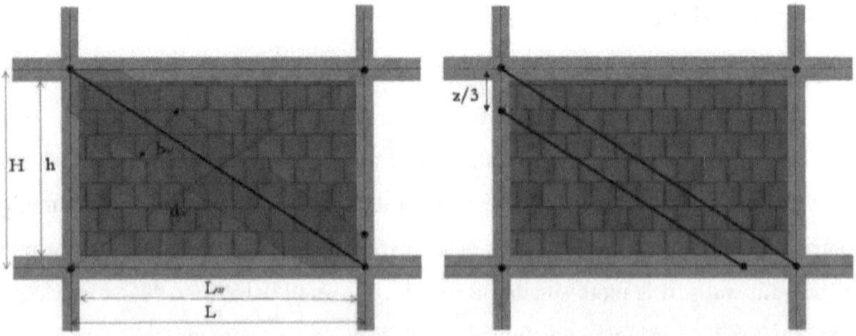

(a) (b)

Figure 1.6 Macro-modelling approaches[87] : (a) Holmes[30] single-rod approach (b) Smith[85] double-rod approach

For the analysis of existing reinforced concrete buildings, often characterised by poor seismic detailing and non-ductile reinforced concrete structural members, it may be more accurate to consider an off-diagonal macro model to simulate possible shear failure mechanisms at the top of columns. For this purpose, more complex macro models have been developed using several diagonal connecting rods. For example, Leuchars and Scrivener[88] proposed a model with two simple connecting rods and a shear spring, located in the centre of the panel, to reproduce the failure of the infill wall by shear slip. In addition, Chrysostomou et al.[89] (Figure 1.7a) and El-Dakhakhni et al.[12] (Figure 1.7b) proposed a six-rod approach, with three rods per direction, to accurately reproduce the local effects due to the interaction between the portal frame and the infill. It should be mentioned that in the case of multi-connecting rod approaches, different prescriptions regarding relative stiffness, inclination and contact length (length of the columns in contact with the side of the infill during horizontal deformation) are provided in the literature and need to be properly calibrated.

As far as relative stiffness is concerned, in the case of three connecting rods, it is generally assumed that 50% and 25% of the total stiffness of the infill wall is attributed to the central and off-diagonal connecting rods respectively. In the case of two rods, if both rods are off-diagonal, the relative stiffness is generally 50% of the total stiffness of the infill wall (for each rod). On the other hand, if only one connecting rod is off-diagonal, the stiffness is generally set at 75% for the diagonal connecting rod and 25% for the off-diagonal connecting rod.

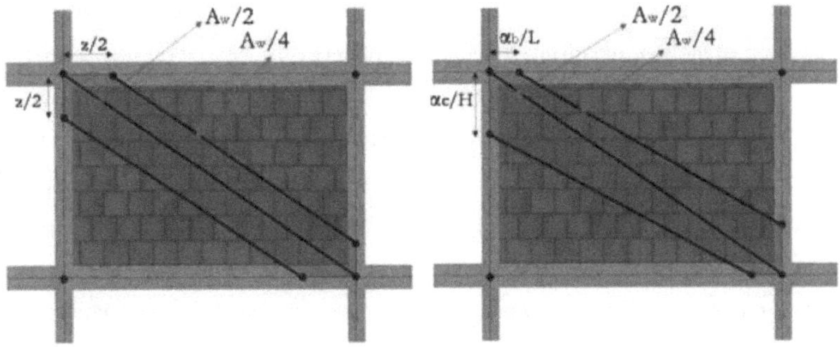

(a) (b)

Figure 1.7 Macro-modelling approaches[87] : (a) Chrysostomou et al.[89] three-rod approach, (b) El-Dakhakhni et al.[12] three-rod approach

Crisafulli[11] has developed a four-node panel element in which compression and shear behaviour are considered separately using a double-rod mechanism and a shear spring in each direction (Figure 1.8a). The model is characterised by internal nodes and dummy nodes, spaced apart according to the contact length; only the internal nodes are connected to the beam-post joints, which remedies the subdivision of the post and more accurately reproduces the local effect due to the interaction between the portal frame and the infill. Finally, another simplified macro model, which is an improvement on the equivalent bi-diagonal compression rod model, has been proposed by Furtado et al.[14] . It consists of four rigid connecting rod elements, which are connected at the beam-post joints, and a central element in which the non-linear behaviour is concentrated (Figure 1.8b).

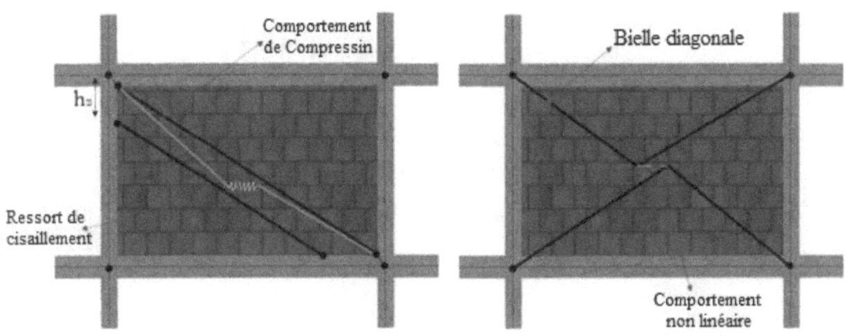

(a) (b)

Figure 1.8 Macro-modelling approaches[87] : (a) Crisafulli[11] macro-model, (b) Furtado et al.[14] macro-model.

Based on this observation, several models have been developed to determine the geometry and strength of the diagonal connecting rod. The mechanical properties required to define the connecting rod depend on the type of analysis carried out (linear or non-linear). For example, for a linear analysis, modulus of elasticity and compressive strength are required. However, for a non-linear analysis, the stress-strain relationship must be defined[90] . In addition, the width of the connecting rod is an essential parameter that must be determined in the geometric properties. In the macro-modelling technique, the thickness and length of the connecting rod are considered identical to those of the infill wall. This approach has evolved from a single-rod model to a multi-rod model, taking into account both the in-plane and out-of-plane behaviour of the wall, as well as their interactions.

1.5. The equivalent diagonal connecting rod model (macro-model)

As mentioned in the previous section, from a practical point of view, the use of continuous modelling approaches (micro-modelling, meso-modelling) is not feasible for large systems due to the high computational cost. Due to their inherent simplicity, rod models (i.e. macro-models) are one of the most practical approaches for representing the behaviour of infill walls, particularly for design and performance evaluation purposes[90] . As shown in Figure 1.9, these modelling approaches involve replacing the infill wall with an equivalent diagonal hinged connecting rod system.

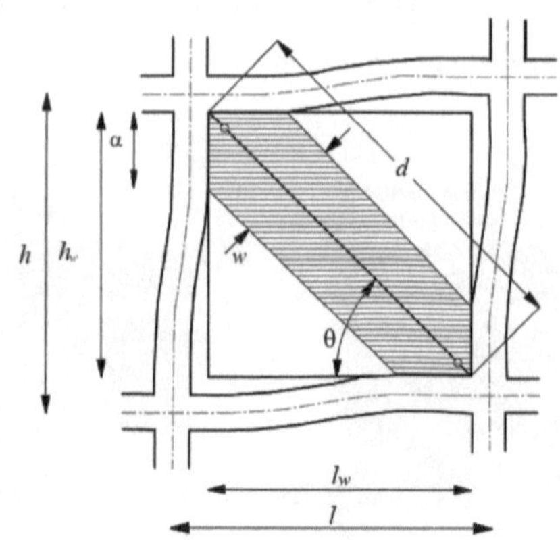

Figure 1.9 The equivalent diagonal connecting rod and its relevant parameters .[90]

Extensive research was carried out to determine the main structural properties of the diagonal connecting rod, such as its width, stiffness, constitutive behaviour and the number of connecting rods to be taken into account. These characteristics vary according to the type of analysis (linear elastic or non-linear) and the type of loading (monotonic, cyclic or transient). For example, for a linear elastic analysis, the properties required for a diagonal connecting rod are its geometric characteristics (length and cross-section) and its modulus of rigidity. However, when the non-linear behaviour of the material is taken into account, the entire stress-strain relationship of the connecting rod is required. In addition, the properties required for the diagonal connecting rod become more complex under cyclic and dynamic loading.

Tucker in 2007[91] has classified the analytical methodologies available for defining in-plane properties of model connecting rods into two main approaches: stiffness methods and strength methods. Both methods replace the infill wall with an equivalent connecting rod, but use different approaches to define the necessary properties of the connecting rod. The stiffness method estimates the structural contribution of the infill wall based on the formation of the compression zone along the infill. Therefore, this method focuses on estimating the geometric properties of the tie-rod and combines these properties with equivalent material properties (typically the compressive strength characteristic of the masonry) to define the lateral capacity of the infill. The resistance method defines the properties of the connecting rod by quantifying the lateral forces supported by the infill wall. A more detailed discussion of these two methods is provided below, including modifications that have been proposed over time to these modelling approaches.

1.5.1. The rigidity method

Polyakov in 1960[84] was the first to suggest the possibility of considering the effect of the infill wall as equivalent diagonal bracing. Subsequently, Holmes in 1961[30] adopted this suggestion and proposed a linear equivalent compression rod model. Holmes replaced the infill wall with a diagonal connecting rod with the same material and thickness as the infill wall, while the width 'w' of the proposed connecting rod was given by the following expression:

$$w = \frac{1}{3}d \qquad (1.1)$$

Where 'd', as shown in Figure 1.10, is the diagonal length of the infill wall. This rule is called the rule of thirds and has been suggested to be applicable irrespective of the relative stiffness of the portal frame and infill. Based on this rule, Holmes proposed expressions for estimating the horizontal displacement of the portal frame with infill at failure and for calculating the horizontal load causing crush failure in the corner regions. A year later, on the basis of experimental data from a large series of tests on steel portal frames with masonry infill,

Stafford Smith[31] in 1962 reported that the "w/d" ratio varied from 0.10 to 0.25 for infill with an "h_w/l_w" side ratio of 5 to 1.

Smith et al.[92] in 1969 further developed the equivalent connecting rod theory by evaluating the interaction between the portal frame and the infill to calculate the effective width of the connecting rod. In their formulation, the effective width of the connecting rod is a function of the relative stiffness of the column and the infill, the length-to-height ratio of the infill, the stress-strain relationship of the infill material and the value of the diagonal load acting on the infill. The contact length "α" between the infill and the portal frame as shown in Figure 1.10 is defined by :

$$\frac{\alpha}{h} = \frac{\pi}{2\lambda h} \quad (1.2)$$

Where "h" is the height of the column between the axes of the beams and "λ" is a characteristic stiffness parameter given by :

$$\lambda = \sqrt[4]{\frac{E_w\, t \sin 2\theta}{4\, EI\, h_w}} \quad (1.3)$$

Where "E_w" is the modulus of elasticity of the masonry wall, "EI" is the flexural stiffness of the columns, "t" is the thickness of the infill wall and equivalent connecting rod, "h_w" is the height of the infill wall and "θ" is the slope of the infill diagonal (in degrees). In addition, they also developed a set of empirical curves that relate the stiffness parameter to the effective width of an equivalent connecting rod.

Based on experimental tests on model portals with two different materials (concrete and brick), Mainstone[33,93] in 1971 and 1974 studied all the variables that could have a major influence on the contribution of infill walls to the resistance of a lateral deformation mechanism in multi-storey buildings. On the basis of this research, he proposed two empirical expressions to evaluate the width of the equivalent connecting rod in the case of brick and concrete infills: equations (1.4) and (1.5) in the case where "λ_h" is between 4 and 5, and equations (1.6) and (1.7) in the case where "λ_h" is greater than 5.

$$w = 0.175\, d\, {\lambda_h}^{-0.4} \quad Brique \quad (1.4)$$

$$w = 0.115\, d\, {\lambda_h}^{-0.4} \quad Béton \quad (1.5)$$

$$w = 0.16\, d\, {\lambda_h}^{-0.3} \quad Brique \quad (1.6)$$

$$w = 0.11\, d\, {\lambda_h}^{-0.3} \quad Béton \quad (1.7)$$

Here, "λ_h" is used to represent the dimensionless term "λ_h" in Eq. (1.2).

Tassios[94] in 1984 simplified the diagram proposed by Bazan and Meli[95] in 1980 to predict the width of the equivalent connecting rod in the case of failure (cracking) along the diagonal of the infill wall by the following expression :

$$w \cong 0.20\ d \sin\theta \sqrt{\frac{E_c A_c}{G_w A_w}} \quad si\ 1 < \frac{E_c A_c}{G_w A_w} < 5 \tag{1.8}$$

Where "G_w" is the shear modulus of the filling. Based on earlier experimental data, Te-Chang and Kwok-Hung[96] in 1984 proposed a semi-empirical expression to calculate the equivalent diagonal connecting rod width when θ is in the range of 25° to 50°, given by :

$$w = \frac{0.95\ h_w \cos\theta}{\sqrt{\lambda_h}} \quad si\ 1 < \frac{h_w}{l_w} < 5 \tag{1.9}$$

Where "$h_w$and l_w" are the height and width of the infill wall respectively.

Considering the state of the infill wall in terms of cracking, Decanini and Fantin[97] in 1986 proposed the following two sets of expressions to quantify the width of the equivalent diagonal link for cracked and uncracked panels described in equations (1.10) and (1.11), respectively :

$$w = \begin{cases} \left(\frac{0.707}{\lambda_h} + 0.01\right) d & si\ \lambda_h \leq 7.85 \\ \left(\frac{0.470}{\lambda_h} + 0.04\right) d & si\ \lambda_h > 7.85 \end{cases} \tag{1.10}$$

$$w = \begin{cases} \left(\frac{0.748}{\lambda_h} + 0.085\right) d & si\ \lambda_h \leq 7.85 \\ \left(\frac{0.393}{\lambda_h} + 0.130\right) d & si\ \lambda_h > 7.85 \end{cases} \tag{1.11}$$

On the basis of experimental results of portal frames with scaled brick infill, Moghaddam[98] in 1988 proposed the following simple relationship between the length of the equivalent diagonal connecting rod and its width:

$$w = \frac{1}{6} d \tag{1.12}$$

Paulay and Priestley[17] in 1992 pointed out that a higher value of connecting rod width will result in a stiffer structure and therefore a higher seismic demand. For seismic design purposes, they recommended using the following expression:

$$w = \frac{1}{4} d \tag{1.13}$$

In the light of previous research and empirical results, Durrani and Luo[99] in 1994 proposed the following semi-empirical expressions to evaluate the width of the equivalent diagonal connecting rod :

$$w = \gamma\, d \sin(2\theta) \tag{1.14}$$

$$\gamma = 0.32\sqrt{\sin(2\theta)}\left[\frac{h^4\, E_w\, t}{m\, E_c\, I_c\, h_w}\right]^{-0.1} \tag{1.15}$$

$$m = 6\left(1 + \frac{6\, E_b\, I_b\, h}{\pi\, E_c\, I_c\, L}\right) \tag{1.16}$$

Where "$E_b$and E_c" are the moduli of elasticity for the beam and column, respectively, while "$I_b$and I_c" are the moments of inertia of the beam and column, respectively.

Flanagan and Bennett[100] in 1999 proposed a fragmented linear connecting rod width calculated by :

$$w = \frac{\pi}{C\, \lambda \cos\theta} \tag{1.17}$$

Where 'C' is an empirical constant based on deformation displacements and infill damage. The values of 'C' vary depending on the type of portal frame and the infill material, according to Bennett et al. .[101]

Al-Chaar[102] in 2002, the equivalent width of the filler rod in the elastic range can be estimated using equations (1.18) to (1.20). It should be noted that linear interpolation is required for aspect ratios between 1.0 and 1.5.

$$w = 0.0835C\, d\left(1 + \frac{2.574}{\lambda_h}\right) \quad si\ l_w/h_w \geq 1.5 \tag{1.18}$$

$$C = -0.3905\left(\frac{l_w}{h_w}\right) + 1.7829 \tag{1.19}$$

$$w = 0.1106\, d\left(1 + \frac{6.027}{\lambda_h}\right) \quad si\ l_w/h_w = 1.0 \tag{1.20}$$

Chethan et al.[103] estimated the equivalent connecting rod width using the contact length between the infill wall, beam and column as follows:

$$w = \sqrt{\alpha_h^2 + \alpha_l^2} \tag{1.21}$$

Where "α_h and α_l" are the vertical and horizontal contact lengths, respectively, given by equations (1.22) and (1.23).

$$\alpha_h = \frac{\pi}{2}\sqrt[4]{\frac{4\,EI_c\,h_w}{E_w\,t\,\sin(2\theta)}} \tag{1.22}$$

$$\alpha_l = \frac{\pi}{2}\sqrt[4]{\frac{4\,EI_b\,l_w}{E_w\,t\,\sin(2\theta)}} \tag{1.23}$$

Where "EI_c and EI_b" in equations (1.22) and (1.23) refer to columns and beams respectively. The other parameters involved are those also defined in Eq. (1.3).

In the same context, Khaja[104] in 2013 reported that "$\alpha_{(l)}$", the contact length between the infill wall and the beam, is approximately equal to half the span of the beam.

Recently, Turgay et al.[105] have proposed alternative expressions to provide improved estimates of the stiffness and deformability of reinforced concrete frames with infill (the "h/l" ratio considered in this study varied from 1.2 to 2). According to their proposal, the width of the diagonal connecting rod is given by the following expression :

$$w = 0.18\frac{d}{\sqrt[4]{\lambda_h}} \tag{1.24}$$

From this review, it is evident that there is significant diversity among the various proposals for defining the properties of the connecting rod model. In the next section, several approaches to defining their constitutive behaviour are presented.

1.5.2. Resistance method

Unlike the stiffness method, the strength method first defines the force-displacement relationship for the filling and then relates this relationship to the area of the connecting rod to define the constitutive behaviour of the filling. A large body of literature presents proposed models for the force-displacement relationship of the equivalent diagonal connecting rod. One of the first attempts to define the complete force-displacement behaviour of the infill wall was made by Klingner and Bertero[35] in 1978. They proposed a non-linear hysteretic response for the equivalent diagonal link model, based on experimental tests carried out on reinforced concrete porticos with masonry infill. They allowed for strength and stiffness degradation during cycling in their model, as shown in Figure 1.10, while the unloading stiffness was assumed to be linear and equal to the initial stiffness. Although their model assumes that the diagonal connecting rod has some tensile strength, they ignored this component in the implementation of the model in their numerical analyses. Experimental calibration of this model showed poor agreement. However, this model was the first attempt to take into account the non-linear cyclic characteristic of the diagonal connecting rod response and represented the basic approach for subsequent developments.

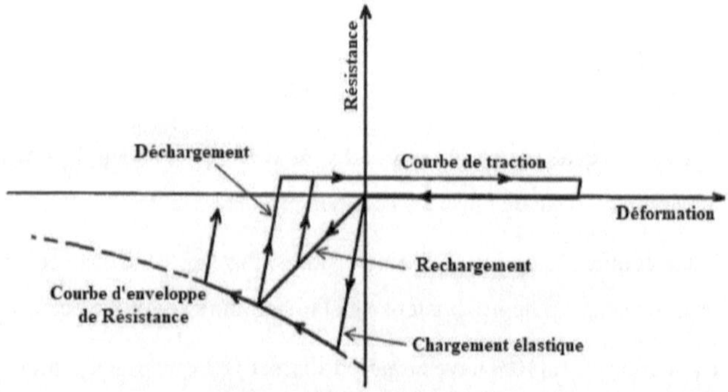

Figure 1.10 The behaviour of the equivalent diagonal connecting rod proposed by Klingner and Bertero .[35]

To define the non-linear behaviour of their proposed model, Andreaus et al.[106] considered the hysteretic curve shown in Figure 1.11. According to this model, strength degradation begins immediately after the connecting rod force reaches its maximum value and reloading occurs when the axial deformation equals the plastic deformation of the previous loop. In addition, their model does not take into account the effects of stiffness degradation.

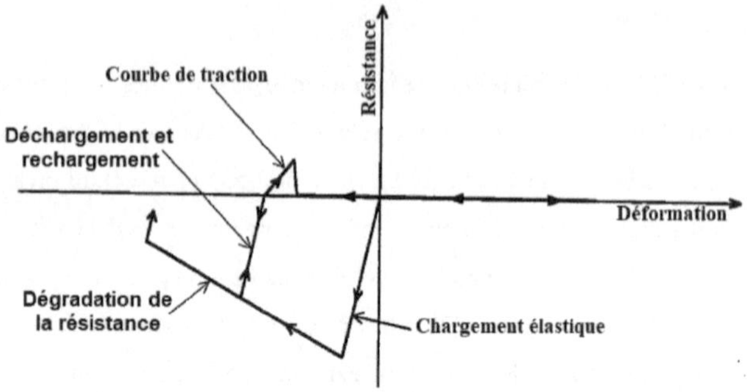

Figure 1.11 The hysteretic model used by Andreaus et al. [106]

Doudoumis and Mitsopoulou[107] in 1986 also proposed a hysteretic model illustrated in Figure 1.12. This model was developed for porticos with non-integral infill, in which a gap is normally formed between the masonry wall and the portico. The envelope curve has taken into account strength degradation, but the hysteretic cycles have been described in a very simplistic way by assuming that reloading occurs along the elastic branch.

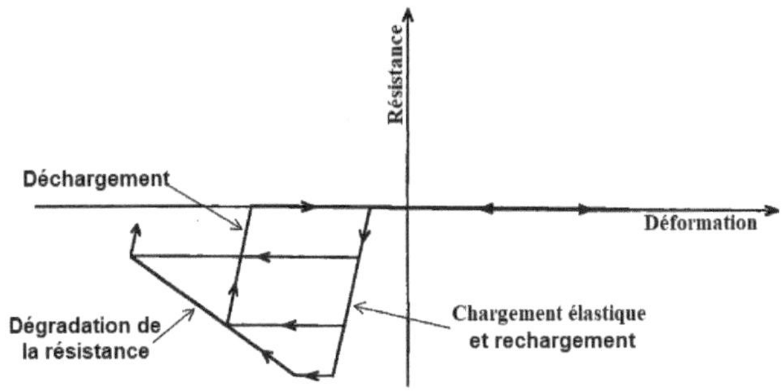

Figure 1.12 The hysteretic model for gantries with non-integral filling by Doudoumis and Mitsopoulou [107]

Madan et al.[108] have combined a set of mathematical functions to develop the smooth force-strain relationship shown in Figure 1.13. By selecting suitable values of the nine parameters included in the model, their force-deformation hysteresis model takes into account the degradation of strength and stiffness, as well as the pinching effect resulting from the opening and closing of the masonry spaces in the infill wall. It should be noted that some of these parameters are empirical, while others depend on energy dissipation considerations. The implementation of this model is not straightforward and requires the numerical integration of a differential equation.

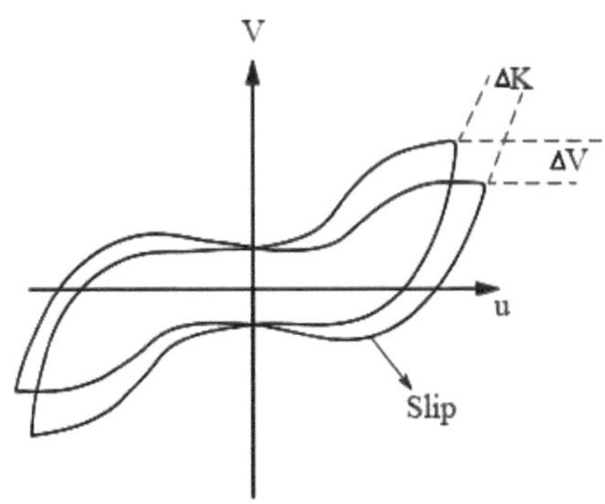

Figure 1.13 Integrated hysteretic model according to Madan et al. .[108]

Based on the simplified stress-strain relationship of the masonry material, El-Dakhakhni et al.[12] used a trilinear response model, as shown in Figure 1.14, in their model of three diagonal connecting rods. Their scheme includes an elastic, plastic (ultimate strength) and post-capping branch for the envelope of connecting rod behaviour.

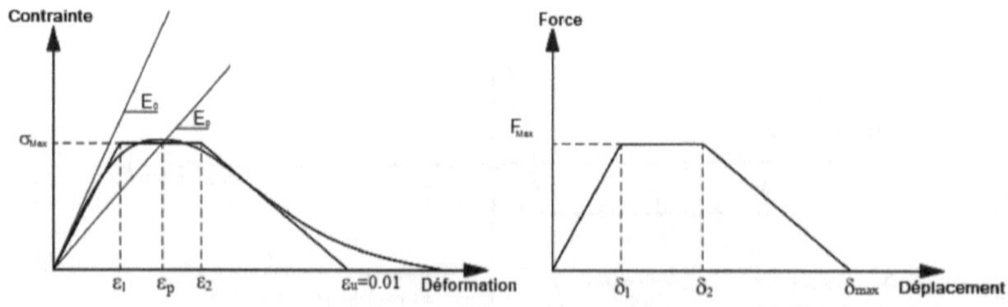

(a) (b)

Figure 1.14 The simplified trilinear relationship of the connecting rod proposed by El-Dakhakhni et al.[12] : a) masonry stress-strain relationship; b) typical force-strain relationship for the connecting rod model.

In the context of using a trilinear relationship to represent the non-linear behaviour of masonry infills, several expressions have been proposed, such as those of Bertoldi et al.[109] , Panagiotakos et al.[110] , Dolšek and Fajfar[10] , Uva et al.[111] or De Risi et al.[112] . As illustrated in Figure 1.15, this trilinear relationship describes the evolution of the lateral force supported by the filling and the corresponding displacement. Therefore, to define the curve, the coordinates of the three main points in terms of the corresponding forces and displacements are required. The expressions available to define these trilinear relationships are either based on regression analysis using experimental data (for example, see Bertoldi et al.[109] and Panagiotakos et al.[110]), or adopt and combine pre-existing expressions available in the literature to obtain a more realistic curve (for example, see Dolšek and Fajfar[10] , Uva et al.[111] and De Risi et al.).[112]

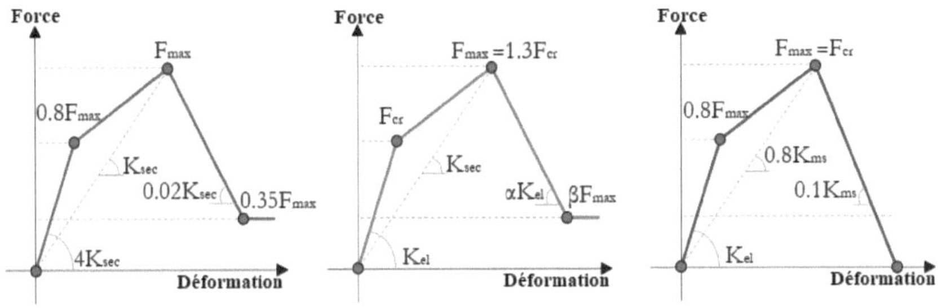

Figure 1.15 Example of trilinear curves[87] : (a) Bertoldi et al.[109] , (b) Panagiotakos and Fardis[110] and (c) De Risi et al. .[112]

In general, there are two distinct approaches to plasticity (the inelastic approach and the concentrated plasticity approach). The distributed inelastic approach involves the propagation of inelastic deformation along the entire length of a member, with its cross-section divided into fibres. This approach is time-consuming but provides a realistic estimate of the distribution of structural damage. On the other hand, the concentrated plasticity approach refers to inelastic deformation occurring at critical sections along a member and is often considered to be more computationally efficient. However, this method requires complicated calculations to determine moment-rotation curves and interaction curves for hinges[113] .

Recently, there has been a growing interest in the use of fibre hinges in the non-linear analysis of structures with infill walls[113–120] . However, their adoption in reinforced concrete portal frames is still rare despite the guidance provided by the Applied Technology Council[121] and their implementation in analysis software such as ETABS and SAP2000.

Di Trapani et al.[119] proposed a method for simulating infill walls using a fibre section model that defines the behaviour of equivalent connecting rods in compression (Figure 1.16), after carrying out validation tests using different experimental results. The results demonstrated the predictive reliability of the proposed approach as an evaluation tool and highlighted the potential of correlation equations as design tools for performance-based design of portal frames with infills.

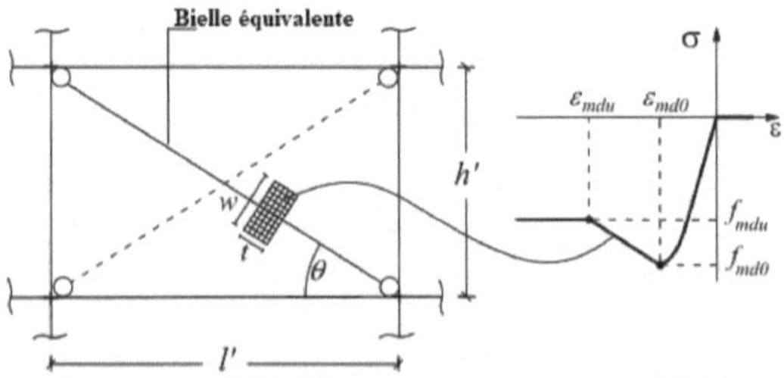

Figure 1.16 Equivalent diagonal connecting rod model with a fibre-discretised section proposed by Di Trapani et al. [119]

Van et al.[113] proposed a macro-modelling method that incorporates a plastic fibre ball joint to simulate the hysteretic response of reinforced concrete portal frames with infill walls using SAP2000 software. By comparing the numerical results with experimental observations, they concluded that the proposed numerical modelling can reasonably predict the overall response of portal frames with in-plane infill walls.

According to Adibi et al.[120] , who used a macro-modelling method incorporating a plastic fibre ball joint from Di Trapani et al.[119] using ETABS software, infill walls in all spans of the precast concrete portal frame reduce the probability of exceeding damage limit states in the structures. They also found that the absence of infill walls on the first floor significantly affects the reliability and seismic performance of the structure.

1.6. Conclusion

From the literature review presented, it is clear that representing the structural behaviour of reinforced concrete frames with masonry walls is complex due to the large number of parameters and phenomena involved, as well as the considerable uncertainty associated with many of these parameters. However, experimental tests on reinforced concrete frames with masonry walls provide important results for understanding and analysing the complex behaviour of these structures under seismic loads and, consequently, validating the structural parameters of the macro-models (connecting rod models). The experimental tests were carried out taking into account several structural aspects such as the material of the infills, the configurations of the infill walls (with or without openings, of different sizes and positions) and the type of loading (monotonic or cyclic loading). However, due to the significant variation in material properties, construction techniques and geometry, the experimental approach is unable to simulate all possibilities because of the high cost of such procedures. In conclusion, experimental and numerical studies of masonry wall structures provide crucial information for understanding their complex behaviour under seismic loading. Although the experimental approach has limitations due to the diversity of structural characteristics, connecting rod models, in particular the representation by a single equivalent diagonal connecting rod with a fibre-discretised cross-section, have proved useful for validating structural parameters in behaviour analyses of reinforced concrete structures with masonry walls.

Chapter 2: Seismic Analysis Methods

2.1 Introduction

To assess seismic performance, a structural analysis of the mathematical model of the structure is required to determine the forces and displacements required at the various levels of the construction. Several analysis methods, both linear and non-linear, are available to predict the seismic behaviour of structures. Consequently, four main types of analysis methods are available:

- Linear analysis methods :
  - Linear statics
  - Linear dynamics
- Non-linear analysis methods :
  - Non-linear statics
  - Non-linear dynamics

2.2 Linear methods

2.2.1. The linear static method

The linear static method uses force-based assessment methodology and is the oldest and simplest method of structural analysis. It assumes that the fundamental mode of vibration controls the structural behaviour. The horizontal distribution of statically applied loads must be close to the first mode, which represents a significant simplification. An equivalent static analysis can assess seismic structural performance. However, these analysis procedures are only appropriate for medium and regular height buildings where the effects of higher modes are not significant . [122]

According to FEMA 356[86] , structures are analysed and evaluated using this method with linearly elastic damping and stiffness values at or near the plasticity level.

According to the RPA 2024 code[123] , the total seismic force applied to the base of the structure is expressed by a mathematical formula in the equivalent static method. The equation. (2.1) :

$$V = \lambda \frac{S_{ad}}{g} (T_0)\, W \tag{2.1}$$

Where:

λ Correction coefficient.

$\frac{S_{ad}}{g}(T_0)$: Ordonnée du spectre de calcul pour la période T_0 .

T_0 Fundamental period of vibration of the building.

W: Total seismic weight of the building.

The empirical formulae used to estimate the seismic force and the distribution of the total seismic force according to the height of the building and the fundamental period are described in equations (2.2) and (2.3), respectively.

$$F_i = \frac{(V - F_t)W_i\, h_i}{\sum_{j=1}^{n} W_j\, h_j} \qquad (2.2)$$

Where:

F_i: Horizontal force returning to level i.

F_t: Concentrated force at the top of the structure

h_i: Floor level where force $F_i$is exerted.

h_j: Any floor level.

W_i, W_j: Weights attributable to floors i, j.

$$T = C_T {h_n}^{3/4} \qquad (2.3)$$

Where:

C_T: Coefficient depending on the bracing system and the type of infill.

h_n: Height measured in metres from the base of the structure to the top floor.

- ***Advantages and disadvantages***
- *Benefits*
- ✓ It is very simple compared with other methods.
- ✓ It gives a primary estimate of the dimensions of the structure.
- ✓ It takes account of plasticity through the behaviour coefficient.

- *Disadvantages*
- ✓ It does not take into account the dynamic behaviour of structures.

- ✓ It is limited for a certain type of building by conditions (irregularity, zones).
- ✓ It takes account of the first mode only, but in reality there are several modes of vibration.

This method simulates the behaviour of reinforced concrete structures and does not take dynamic behaviour into account. In addition, it is subject to several application conditions.

2.2.2. The linear dynamic method

When the contribution of the higher modes to the structural response is significant, linear dynamic procedures are appropriate methods and their results are more accurate than those of linear static procedures.

According to FEMA 356[86] , the linear method should be used when buildings are modelled with equivalent viscous damping values and linearly elastic stiffness at the plasticity level of this method. FEMA 356[86] suggests the modal spectral method and temporal dynamic analysis. In these methods, a linear elastic analysis based on the superposition principle is used to obtain the internal displacements and forces of the system.

This method is widely recognised as being effective for calculating the linear dynamic response of elastically damped systems. The method is interesting because the response of multi-degree-of-freedom systems is expressed by a modal superposition, with the response of each mode determined from the spectral analysis of single-degree-of-freedom systems. The modal analysis combines the responses of the different modes using the superposition rules to obtain the mean value of the maximum structural response.

Response spectrum method

Systems with multiple degrees of freedom (multi-ddl) are usually analysed using modal analysis. A typical multi-ddl system with 'n' degrees of freedom is shown in Figure 2.1. This system, when subjected to ground motion, undergoes deformation in many possible ways. These deformed shapes are called modes of vibration or mode shapes. Each shape vibrates with a particular natural frequency. The total number of unique modes for each multi-ddl system is equal to the degree of freedom possible for the system. The equations of motion for the multi-ddl system are given by[124] :

$$[M]\{\ddot{x}(t)\} + [C]\{\dot{x}(t)\} + [K]\{x(t)\} = -[M]\{r\}\{\ddot{x}_g(t)\} \quad (2.4)$$

Where:

[M] = Mass matrix (n × n).

[K] = Stiffness matrix (n × n).

[C] = Damping matrix (n × n).

{r} = Influence coefficient vector (n × 1).

{x(t)} = Relative displacement vector.

$\{\dot{x}(t)\}$ = Relative speed vector.

$\{\ddot{x}(t)\}$= Relative acceleration vector.

$\ddot{x}_g(t)$= Seismic ground acceleration.

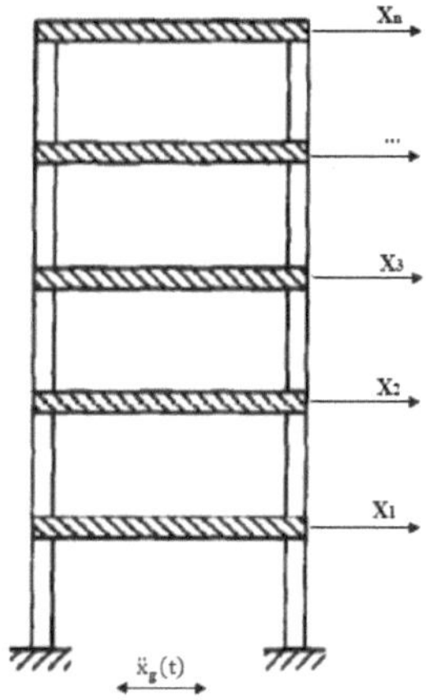

Figure 2.1 Multi-ddl system with "n" degrees of freedom

The eigenvectors of the multi-ddl system can be found in the following equation :

$$\{[K] - \omega_i^2[M]\}\{\emptyset_i\} = 0 \rightarrow det\left|\{[K] - \omega_i^2[M]\}\right| = 0 \qquad (2.5)$$

Where:

ω_i^2 = Eigenvalues of the i^th^mode.

$\{\emptyset_i\}$ = Eigenvector or mode form of the i^{th}mode.

ω_i = Natural frequency in the i^{th}mode.

Or the displacement response of the system at several ddl is expressed as

$$\{x(t)\} = [\emptyset]\{y(t)\} \qquad (2.6)$$

Where,{y(t)} represents the modal displacement vector, and[Ø] is the mode shape matrix given by :

$$[\emptyset] = [\emptyset_1, \emptyset_2, \dots \dots \dots, \emptyset_n] \tag{2.7}$$

ReplaceEq. (2.6) in Eq. (2.4) and multiply by$[\emptyset]^T$:

$$[\emptyset]^T[M][\emptyset]\{\ddot{y}(t)\} + [\emptyset]^T[C][\emptyset]\{\dot{y}(t)\} + [\emptyset]^T[K][\emptyset]\{y(t)\} = -[\emptyset]^T[M]\{r\}\{\ddot{x}_g(t)\}$$

$$[M^*]\{\ddot{y}(t)\} + [C^*]\{\dot{y}(t)\} + [K^*]\{y(t)\} = -[\emptyset]^T[M]\{r\}\{\ddot{x}_g(t)\} \tag{2.8}$$

Where,

$[M^*]$ = Generalised mass matrix.

$[C^*]$ = Generalised damping matrix.

$[K^*]$ = Generalised rigidity matrix.

Due to the properties of[Ø] , the matrices$[M^*]$ and$[K^*]$ are diagonal matrices. However, for the classically damped system (i.e. if$[C^*]$ is also a diagonal matrix), Eq (2.8) reduces to the following equation:

$$\ddot{y}_i(t) + 2\xi_i\omega_i\,\dot{y}_i(t) + \omega_i^2\,y_i(t) = -\Gamma_i\,\ddot{x}_g(t) \tag{2.9}$$

Where,

$y_i(t)$ = Modal shift response in the ithmode.

ξ_i = Modal damping ratio in the ithmode.

Γ_i = Modal participation factor for the ithmode expressed by :

$$\Gamma_i = \frac{\{\emptyset_i\}^T[M]\{r_i\}}{\{\emptyset_i\}^T[M]\{\emptyset_i\}} \tag{2.10}$$

The maximum modal shift response is found from the response spectrum, i.e. :

$$y_{i,max} = |y_i(t)|_{max} = \Gamma_i\,S_d(\xi_i\,,\omega_i) \tag{2.11}$$

Where$S_d(\xi_i\,,\omega_i)$ is the relative displacement spectrum of the seismic ground motion for the parametersξ_i and .ω_i

The maximum displacement response of the structure in the ith mode is :

$$x_{i,max} = \emptyset_i\,y_{i,max} \tag{2.12}$$

The maximum acceleration response of the structure in the ithmode is :

$$\{\ddot{x}_a\}_{i,max} = \{\emptyset_i\}\,\Gamma_i\, S_{pa}(\xi_i\,,\omega_i) \tag{2.13}$$

Where,$S_{pa}(\xi_i\,,\omega_i)$ is the pseudo-acceleration response, obtained by multiplyingS_d to$\omega_i{}^2$, so :

$$S_{pa}(\xi_i\,,\omega_i) = \omega_i{}^2\, S_d(\xi_i\,,\omega_i) \tag{2.14}$$

The required amount of response of interest, "$r_{(i)}$"i.e. (displacement, shear force, bending moment, etc.) of the structure can be obtained in each vibration mode using the maximum response obtained in equations (2.13) and (2.14). However, the final maximum response, "$r_{(max)}$", must be obtained by combining the response in each vibration mode using modal combination rules. Some commonly used modal combination rules are described here .[124]

Modal combination rules

Common methods used to obtain the peak response quantity of interest for a multi-ddl system are[125] :

- Absolute sum method (ABSSUM).
- Square root sum of squares (SRSS) method.
- Complete quadratic combination (CQC) method.

In the ABSSUM method, the peak responses of all modes are added algebraically, assuming that all modal peaks occur at the same time. The maximum response is given by[125] :

$$r_{max} = \sum_{i=1}^{n} |r_i| \tag{2.15}$$

The ABSSUM method (Figure 2.2) provides a very conservative estimate of the amount of resulting response and thus provides an upper limit to the maximum value of the total response.

In the SRSS method, the maximum response is obtained as the square root of the sum of the squared responses in each vibration mode, expressed as[125] :

$$r_{max} = \sqrt{\sum_{i=1}^{n} r_i{}^2} \tag{2.16}$$

The SRSS method of combining the maximum modal responses is basically sound when the modal frequencies are well separated. However, this method gives poor results when the frequencies of the main contributing modes are very close to each other .[125]

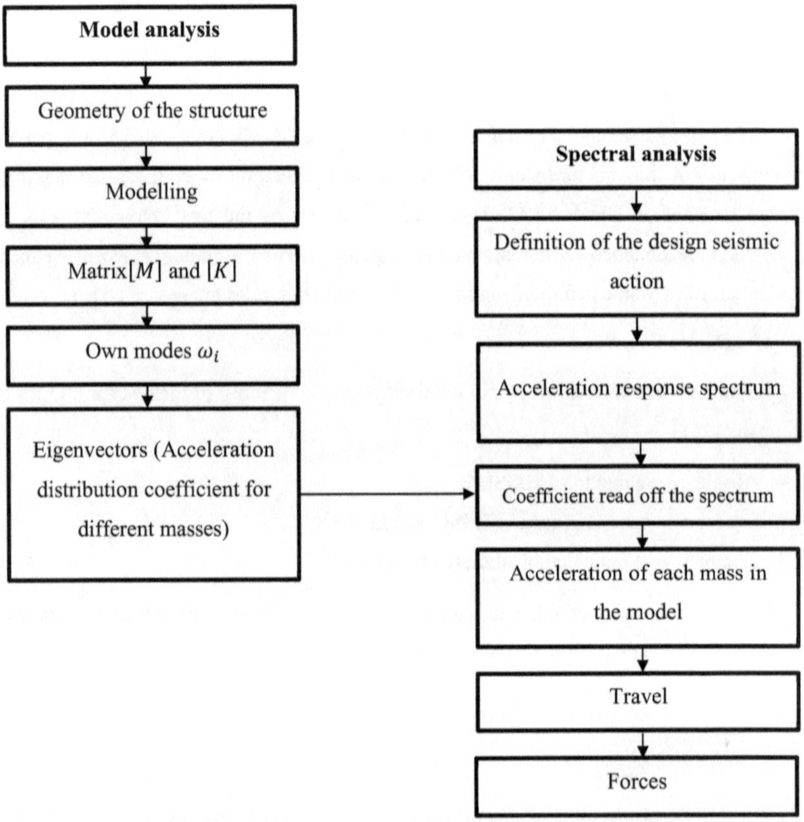

Figure 2.2 Flow chart of the spectral modal method [124]

The alternative procedure is the full quadratic combination (CQC) method. The maximum response of all modes is calculated as follows[125] :

$$r_{max} = \sqrt{\sum_{i=1}^{n} \sum_{j=1}^{n} r_i \, \alpha_{ij} \, r_j} \tag{2.17}$$

Where, "r_i and r_j" are maximum responses in the ith and j^{th} modes, respectively and "α_{ij} " is the correlation coefficient given by :

$$\alpha_{ij} = \frac{8(\xi_i\xi_j)^{1/2}(\xi_i + \beta\xi_j)\beta^{3/2}}{(1-\beta^2)^2 + 4\xi_i\xi_j\beta(1+\beta^2)^2 + 4(\xi_i^2 + \xi_j^2)\beta^2} \quad (2.18)$$

Whereξ_i andξ_j are the damping ratio in the ithand jthmodes of vibration, respectively.

$$\beta = \frac{\omega_i}{\omega_j} \quad avec: \omega_j > \omega_i \quad (2.19)$$

For the system with the same damping ratio in two modes, i.e.$\xi_i = \xi_i = \xi$, then

$$\alpha_{ij} = \frac{8\xi^2(1+\beta)\beta^{3/2}}{(1-\beta^2)^2 + 4\xi^2\beta(1+\beta^2)^2 + 8\xi^2\beta^2} \quad (2.20)$$

- ***Advantages and disadvantages***

- *Benefits*
  - ✓ It takes account of seismic loads, which are dynamic in nature.
  - ✓ It takes account of other modes of vibration.
  - ✓ It applies to both regular and irregular structures.
- *Disadvantages*
  - ✓ It does not take plasticity into account, except through the behaviour coefficient, which is a coefficient that is estimated according to the type of bracing.

The spectral modal method takes account of ductility by means of a coefficient called the behaviour coefficient R, which is fixed for each type of structure. In addition, it cannot give us the weak points of portal frame structures. In an attempt to overcome the shortcomings of these two linear methods, a new non-linear method was used, which will be explained in more detail in the next section.

2.3 Non-linear methods

The most accurate seismic analysis method for structures is non-linear dynamic analysis, which uses earthquake recordings to obtain the necessary evaluation parameters. However, this can be time-consuming and economically unjustifiable, particularly for structures with 3D models[126] . In addition, to increase accuracy, several seismic recordings with similar site conditions need to be used, further complicating the analysis process.

Engineers consider that the use of non-linear static analysis methods in structural analysis is important and unavoidable in order to reduce the problems of linear dynamic analysis. Compared with traditional analysis, this type of study helps in the interpretation of results and gives more details about the seismic behaviour of a structure. In addition, the development of non-linear computer software and the concept of performance-based engineering have led to the increasing use of non-linear analyses, such as pushover analysis.

Current seismic design standards stipulate that the elastic response is only accurate for small amplitudes. To better approximate real behaviour, it is essential to consider the response of structures beyond the linear elastic domain, by taking plastic behaviour into account. Furthermore, assessing the seismic vulnerability of existing structures using linear methods, as discussed in the first section, generally leads to a less faithful representation of physical reality, and often in an unfavourable direction. Thus, the study of the performance of reinforced concrete structures with masonry infill walls will be carried out by means of a non-linear static analysis of the "pushover" type, a method that is becoming increasingly widespread due to its simplicity for assessing the seismic response of structures while taking into account the non-linear characteristics of the materials.

2.3.1. The non-linear static analysis method

The non-linear static pushover analysis method is an approximate approach in which the structure is subjected to an increasing lateral load until it reaches a target displacement[127] . The non-linear pushover calculation is an interesting alternative, as it shares similarities with non-linear time-domain analysis in terms of material data, while avoiding the complexities associated with step-by-step calculation in the time domain .[128]

The "Pushover" method is based on the principle that the capacity to achieve a level of performance is assessed in terms of displacements. The structure is subjected to incremental loading distributed along its height, and the capacity curve (Figure 2.3) is obtained by controlling the displacement at the head of the structure .[129]

The load distribution and target displacement are based on the assumption that the response is fundamentally controlled by a single mode of vibration and that the shape of this mode remains constant during the earthquake .[130,131]

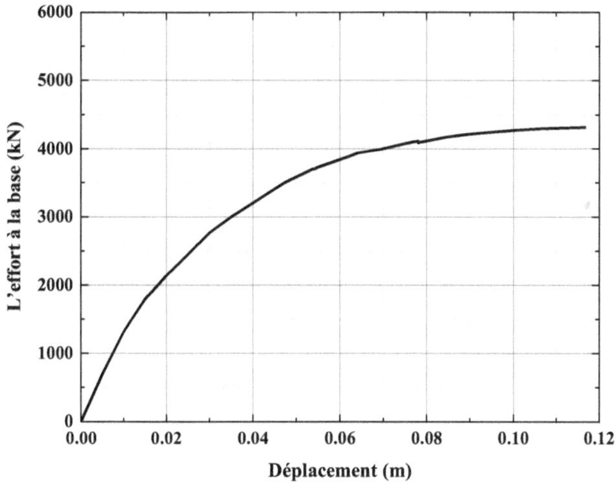

Figure 2.3 Capacity curve of a structure

The principle of the pushover method consists of superimposing a curve representing the resistance capacity of a structure resulting from a non-linear pushover analysis on a curve representing the stress caused by the earthquake (response spectrum) .[129]

The intersection of these two curves, evaluated on the basis of the considerations that follow, represents a point of performance that makes it possible to evaluate the maximum displacement that the structure will undergo and, subsequently, its degree of penetration into the plastic domain.

Two main approaches are proposed, one using damping, as in ATC-40[132] , the other using ductility, as in Eurocode 8 .[133]

Eurocode 8 allows a performance point to be found on the basis of the rule of equal displacements, the inelastic spectrum is obtained by reducing the elastic spectrum by a factor R which depends on the ductility of the structure (μ), while the American standards provide for iterations with several spectra representing different viscous damping coefficients .[134]

2.3.2. Lateral loading mode used in the non-linear static analysis

The lateral loading mode used in the non-linear static analysis is one of the factors that influence the results of the analysis. This is because it represents the distribution of inertial forces generated in the elements of the structure studied when it is subjected to a seismic shock[135] . This distribution varies according to the intensity and duration of the shock. Determining this distribution precisely is complex, as it is linked to the dynamic characteristics of the structure, the free vibration modes, and the degree of participation of the free vibration modes in the overall response. In addition, this distribution varies when the structure is subjected to strong seismic shocks, which can lead to the elastic limit of certain elements being exceeded, thus modifying the distribution of rigidities and internal forces, and increasing the periods of vibration[136,137] . For this reason, several lateral loading modes have been proposed. Initially, they were of a linear and constant nature, mainly used during the design phases, and integrated into traditional analysis methods, notably the so-called "pushover" analysis methods[138,139] . The main modes are as follows:

A. Code side-loading mode

For the non-linear static analysis, ATC-40[132] recommends the use of a static load with a triangular distribution similar to that used in equivalent linear static methods, in particular that mentioned in the Algerian seismic code RPA 2024. The triangular distribution mentioned in the RPA 2024 code is described by the following relationship:

$$F_i = \frac{(V - F_t)W_i\, h_i}{\sum_{j=1}^{n} W_j\, h_j} \tag{2.21}$$

B. Uniform loading mode

The lateral force at each level is proportional to the mass of that level, in another sense

:

$$F_i = \frac{m_i}{\sum m_i} \tag{2.22}$$

Where:

F_i: The lateral force at level i.

m_i: The mass at level i.

These modes give good results for medium-height and lower-height structures. But for high structures, their responses will include: the participation of the higher mode of vibration and the change in dynamic characteristics during non-linear behaviour.

C. **Loading mode according to basic mode**

This vertical distribution is directly linked to the shape of the fundamental mode in the direction taken into account in the analysis. Its use is only authorised when more than 75% of the total mass of the structure is involved in this mode.

$$F_i = \frac{W_i \, \emptyset_i}{\sum_{j=1}^{n} W_j \, \emptyset_j} \tag{2.23}$$

Where:

W_i: The weight of level i.

$\emptyset_j$ The amplitude of level j in the first mode.

2.3.3. Performance levels

Before starting the non-linear static analysis of the structure, an essential step is to determine the level of performance required of the structure when subjected to a specific seismic threat, and to characterise the permissible damage, both to structural and non-structural elements.

The performance level is defined in accordance with the ATC-40 standard[132] as the state of the structure studied after being affected by an earthquake of a given level. In other words, the performance level represents the upper limit of tolerable damage in a structure following its exposure to a certain level of seismic threat. The codes establish two classifications for performance levels: one for structural performance and the other for non-structural performance.

Structural performance levels are defined according to FEMA 356[86] as follows (Figure 2.4):

A. **Level IO (start of operations)**

This level of performance indicates that damage to the building after an earthquake is very limited, and the building's horizontal and vertical strength systems retain their strength and rigidity almost as they did before the earthquake. The risk to life from structural damage is very low, but some simple structural repairs may still be required.

B. **Level LS (safe operating condition)**

This indicates that the damage sustained by the building after an earthquake is significant, but that there is still a margin of resistance to collapse. Some structural elements and components are severely damaged, and there is a significant amount of debris both inside and outside the building. The damage is not significantly life-threatening during the earthquake, but use of the building may be prohibited until repairs are carried out.

C. CP level (damage condition)

This indicates that the building is at risk of partial or total collapse due to significant damage to both structural and non-structural elements. There is a high probability of severe degradation of the stiffness of the lateral load resisting systems, with only a small margin of resistance to collapse. At this level, the main elements of the lateral load resisting systems must continue to resist gravitational forces, as there is a high risk of falling structural debris. It may not be technically feasible to repair the structure safely, and it is unusable due to the presence of aftershocks, which could cause the building to collapse.

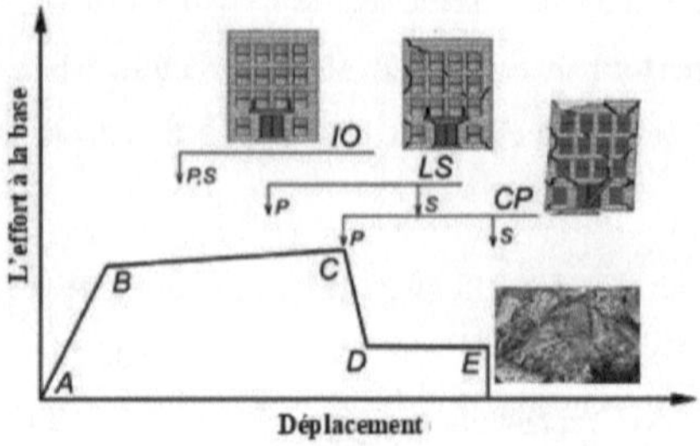

Figure 2.4 Performance levels according to FEMA 356 [86]

2.3.4. Determining the performance point (Target displacement)

Once the lateral forces have been applied to the structure and the capacity curve has been obtained, it is necessary to determine the performance point, corresponding to the target displacement. This target displacement is achieved by the structure when subjected to a specific seismic excitation, often represented by a design response spectrum related to the area under study. Several methods are used to determine this target displacement, in accordance with non-linear analysis codes. The most important methods are the N2 method, the capacitance spectrum method and the displacement coefficient method][128 . The N2 method was introduced in the new RPA 2024 rules .[123]

2.3.4.1. The N2 method

The N2 method, presented by Fajfar in 1999[140] for rapid seismic evaluation of constructions and incorporated in Eurocode 8[133] , has recently been extended to include frames with infill walls[141,10] . It is based on a portal frame model with storeys that are rigid in their planes, generally neglecting vertical displacements, and considering only the horizontal components of seismic motion, namely x and y.

- **One-degree-of-freedom systems and the capacity curve in Eurocode 8**

The N2 method transforms the response of a multi-degree-of-freedom (multi-ddl) system into the response of an equivalent single-degree-of-freedom (one-degree-of-freedom) system. This transformation is necessary to compare the building capacity curve with the demand, as expressed in the design codes by the design spectra, which refer to one-degree-of-freedom systems (Figure 2.5).

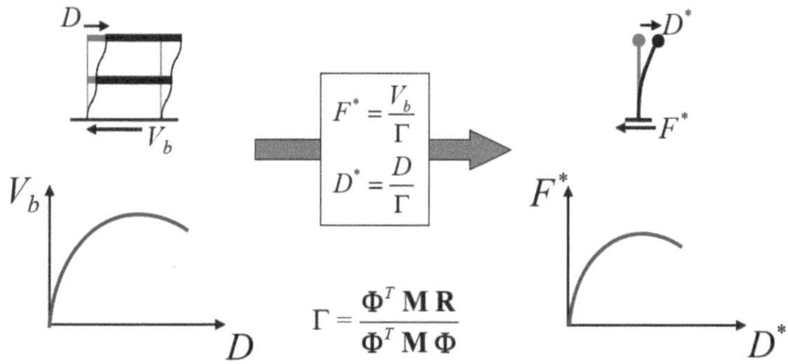

Figure 2.5 The capacity curve: transforming the response of a multi-degree-of-freedom system into a single-degree-of-freedom system [142]

- **Linearisation of the capacity curve and comparison with the demand spectrum**

To compare the capacity curve with the demand curve provided by the design spectrum, the non-linear thrust curves of one-degree-of-freedom systems are approximated by perfectly plastic (or bilinear) elastic curves. This can be based on the principle of energy equality. Assume a target displacement, and assume energy equality between the bilinear and non-linear thrust curves. This simple procedure is illustrated in Figure 2.6.

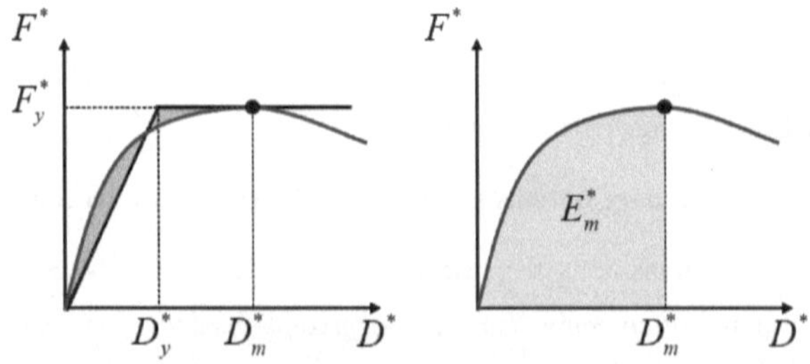

Figure 2.6 The bilinearisation of the capacitance curve of a system with one degree of freedom [142]

The bilinearisation in Figure 2.6 gives the yield point force (F_y^*) and the yield point displacement (D_y^*).

$$D_y^* = 2\left(D_m^* - \frac{E_m^*}{F_y^*}\right) \tag{2.24}$$

This allows us to calculate the initial elastic period as follows:

$$T^* = 2\pi\sqrt{\frac{m^*\,D_m^*}{F_y^*}} \tag{2.25}$$

Secondly, the capacitance curve is transformed into a capacitance spectrum by normalising the force with respect to the weight of the single degree of freedom (SDOF) system. The resulting capacitance spectrum is shown in Figure 2.7.

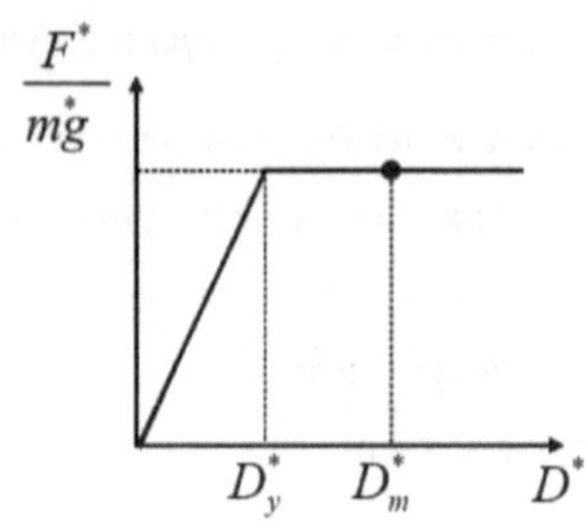

Figure 2.7 The capacitance spectrum of a system with one degree of freedom [142]

➢ **Seismic demand**

The demand on the building is given by the design spectrum provided by the design codes. To compare capacity and demand, the first step is to transform the format of the design spectrum from the traditional Acceleration vs Period format to the Acceleration vs Displacement (ADRS) format. The procedure is fairly straightforward, since acceleration and displacement are linked by the following relationship:

$$S_D = \left(\frac{T}{2\pi}\right)^2 S_A \qquad (2.26)$$

The transformation to the ADRS spectrum is shown in Figure 2.8. The lines from the origin represent constant periods.

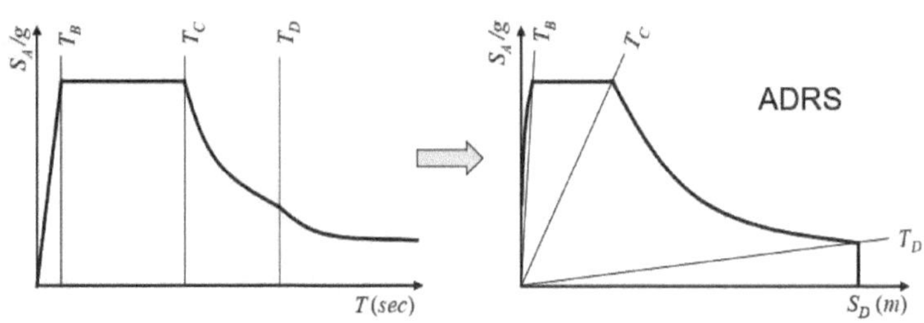

Figure 2.8 The transformation to ADRS spectrum [142]

The capacity spectrum in Figure 2.7 is now compared with the ADRS demand spectrum in Figure 2.8. The comparison is not immediate, as the capacity spectrum is non-linear, whereas the ADRS spectrum provided by the design codes is linear.

For a one-degree-of-freedom system with bilinear plastic behaviour, the acceleration spectrum $S_A$and the displacement spectrum $S_D$can be determined as follows:

$$S_A = \frac{S_{Ae}}{R_\mu} \qquad (2.27)$$

$$S_D = \frac{\mu}{R_\mu} S_{De} = \frac{\mu}{R_\mu}\frac{T^2}{4\pi^2} S_{Ae} = \mu \frac{T^2}{4\pi^2} S_A \qquad (2.28)$$

Where the sub-index "e" indicates elasticity, μ is the ductility factor equal to the maximum inelastic displacement/elastic limit displacement, and Rμ is the reduction factor due to ductility. The reduction factor Rμ can be determined in various ways, some analytical, others approximate. In the simplified version of the N2 method, the following approximate expressions are given:

$$R_\mu = (\mu - 1)\frac{T}{T_C} + 1, \qquad T < T_C \qquad (2.29)$$

$$R_\mu = \mu, \qquad T \geq T_C \qquad (2.30)$$

Where Tcis a characteristic period of ground movement which depends on the type of ground.

The above equations suggest that in the short period domain, the principle of equal displacements is applied (elastic and inelastic one-degree-of-freedom systems have the same maximum displacement), while in the medium and long period domain, the principle of equal energy is applied. These principles are illustrated in Figure 2.9.

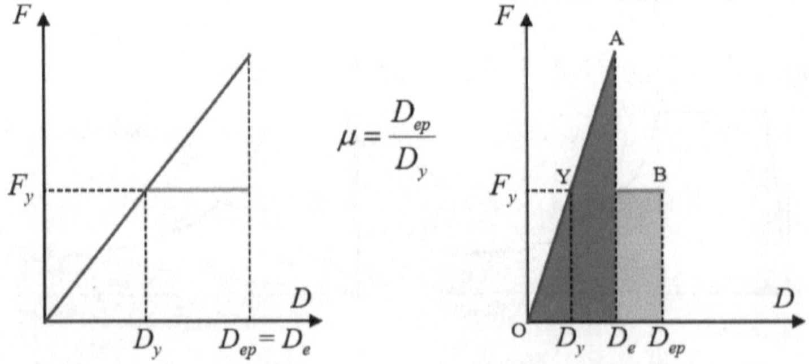

Figure 2.9 The transformation from elastic to bilinear response: assumptions of equality of maximum displacement and equality of energy [142]

The above equations are used to obtain inelastic demand spectra at constant ductility, as shown in Figure 2.9. It should be noted that this procedure is approximate, and that inelastic demand spectra can be determined rigorously (but more complexly) using non-linear dynamic analysis.

The seismic demand on the equivalent of a one-degree-of-freedom system can be determined. The steps are schematically illustrated below for a bilinear oscillator with a short, medium and long elastic period $.T^*$

a) Short elastic period: $.T^* < T_C$

1) $\frac{F_y^*}{m^*} \geq S_A(T^*) \rightarrow la\ réponse\ reste\ élastiquement\ linéaire\ (cas\ a)$

$$D_t^* = D_{et}^* \tag{2.31}$$

2) $\frac{F_y^*}{m^*} < S_A(T^*) \rightarrow la\ réponse\ entre\ dans\ le\ plateau\ non\ linéaire\ (cas\ b)$

$$D_t^* = \frac{D_{et}^*}{q_u\left(1 + (q_u - 1)\frac{T_C}{T^*}\right)} \quad (2.32)$$

Where:q_u is the reduction factor

$$q_u = \frac{S_{Ae}(T^*)}{F_y^*/m^*} \quad (2.33)$$

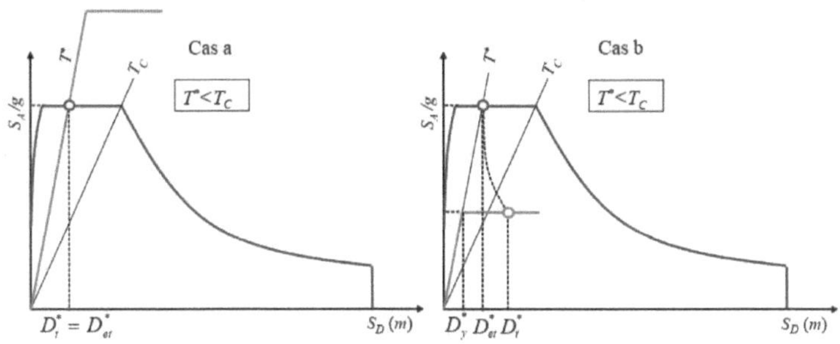

Figure 2.10 Capacity and demand spectra for a short period T^*[142]

b) Medium and long elastic periods: .$T^* \geq T_C$

$$D_t^* = D_{et}^* \quad (2.34)$$

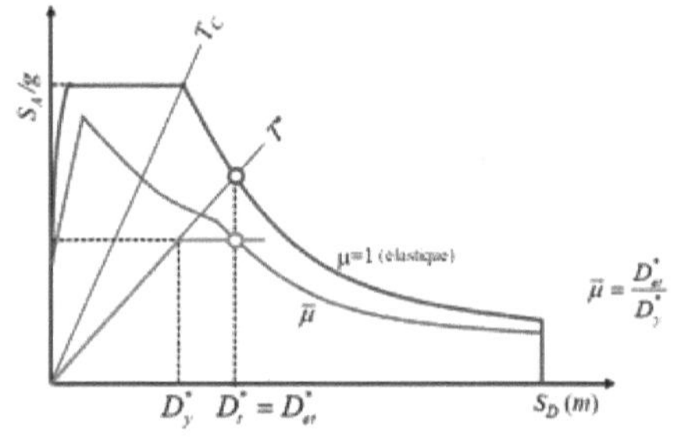

Figure 2.11 Capacity and demand spectra for medium and long period T* with target displacement determination [142]

The target displacement at the top of the building is obtained by reversing the transformation in Figure 2.5, i.e. .$D_t = \Gamma D_t^*$

2.3.4.2. Capacitance spectrum method

The capacity spectrum method, developed by Freeman[143] , is a seismic assessment tool for buildings. The method compares the capacity of the building (the capacity curve) with the requirements on the building (the response spectrum). The non-linear and inelastic behaviour of the structural system is incorporated by applying effective viscous damping values to the linear elastic response spectrum. The intersection of the two curves represents the response of the structure. All the steps in the procedure are described below:

- **Non-linear static analysis (Pushover) of the multi-ddl model**

The structure is pushed by an assumed vertical distribution of lateral load. The assumed shape can be based on the fundamental mode shape. Other lateral force schemes can also be used instead. The capacity curve is then obtained using non-linear static analysis.

- **Definition of the inelastic equivalent one-degree-of-freedom system**

The capacity curve of the multi-ddl system is then approximated by a bilinear curve. The idealised curve, defined by two lines, is obtained by satisfying the following conditions:

The first line is$F = K\delta$, where$F$ is the base force,δ is the roof displacement and$K$ is the initial stiffness; The second line passes through the ultimate capacity point(V_{pi}, μ_{pi}) ; The areas associated with the capacity curve and the bilinear form are the same (the energy equality criterion). Thus, this bilinear capacity curve is defined by the effective deformation point(V_y, μ_y) and the effective inelastic limit (the ultimate capacity point),(V_{pi}, μ_{pi}) . (See Figure 2.12).

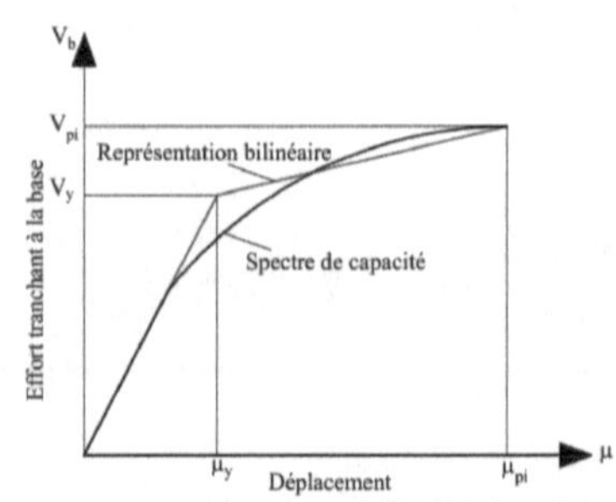

Figure 2.12 Bilinear representation of the capacity curve

- **Converting the capacity curve into a capacity spectrum curve**

The capacitance curve (base force vs. displacement relationship) is then converted to spectral acceleration and spectral displacement, and plotted in an acceleration-displacement response spectrum (ADRS) format. This is based on the following equations:

$$S_a = \frac{V_b}{\alpha_m M} \quad (2.35)$$

$$S_d = \frac{u}{PF_1 \emptyset_{ij}} \quad (2.36)$$

Where, M is the total mass of the building,$\emptyset_{ij}$ is the modal amplitude at floor "i" for mode j,PF_1 the modal participation factor andα_m the modal mass coefficient given by :

$$PF_1 = \frac{\{\emptyset\}[M]\{1\}}{\{\emptyset\}^T[M]\{\emptyset\}} \quad (2.37)$$

$$\alpha_m = \frac{\left[\sum_{j=1}^{n} m_i \phi_{ij}\right]^2}{\sum_{i=1}^{n} m_i \sum_{j=1}^{n} m_i \phi_i^2} \quad (2.38)$$

➢ **Response Spectrum Curve in ADRS format**

The conversion from capacitance curve to capacitance spectrum curve requires the elastic response or design spectrum to be plotted in the acceleration-displacement format, ADRS, rather than the acceleration-period format (see Figure 2.13). The ADRS spectrum is also referred to as the demand spectrum.

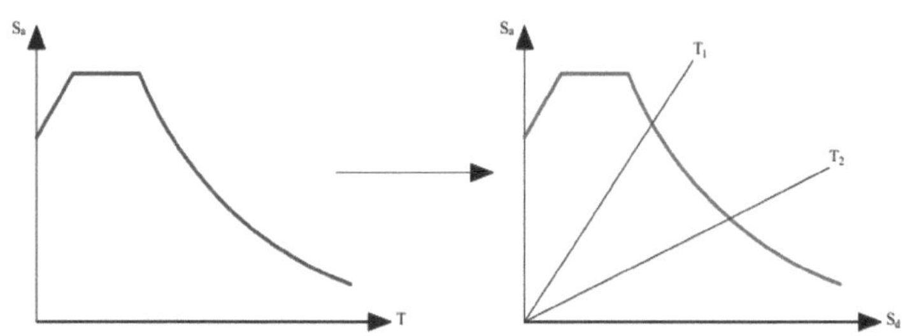

Figure 2.13 Transformation of the elastic response spectrum into an ADRS spectrum

➢ **Superposition of the capacitance spectrum on the damped elastic demand spectrum**

Once the capacitance spectrum and the elastic demand spectrum at 5% damping are plotted on the same curve in the ADRS format (Figure 2.14), an initial estimate of the performance point (a_{pi}, d_{pi}) using the equal displacement rule can be obtained by extending the linear part of the capacitance spectrum until it intersects the elastic demand spectrum at 5% damping. Alternatively, the performance point can be assumed to be the end point of the capacitance spectrum or can be another point chosen based on the engineer's experience, as presented in ATC-40 .[132]

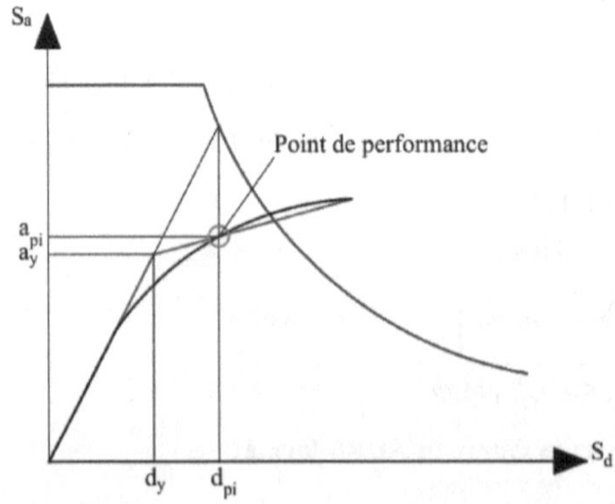

Figure 2.14 The initial estimate of the performance point.

➢ **Equivalent viscous damping**

The damping that occurs when the earthquake pushes the structure into the inelastic domain can be seen as a combination of the viscous damping that exists intrinsically in the structure and hysteretic damping. Hysteretic damping is the damping associated with the surface at the bottom of the loops that form when the seismic force (the shear force at the base) is plotted against the displacement of the structure.

Hysteretic damping can be represented as equivalent viscous damping using the equations available in structural dynamics. The equivalent viscous damping ($B_{(eq)}$), associated with the maximum displacementd_{pi} can be estimated from the following equation:

$$B_{eq} = B_0 + 0.05 \qquad (2.39)$$

B_0: the hysteretic damping represented as equivalent viscous damping 0.05 represents 5% of viscous damping that exists intrinsically in the structure (assumed to be constant), can be calculated as follows:

$$B_0 = \frac{1}{4\pi}\frac{E_D}{E_{S_0}} \qquad (2.40)$$

Where,E_D : energy dissipated by damping;E_{S_0} : maximum deformation energy.

The physical significance of the boundariesE_D andE_{S_0} in equation (2.40) is illustrated in Figure 2.15, whereE_D represents the energy dissipated by the structure in a single motion cycle andE_{S_0} represents the area confined in a single hysteretic loop.E_D is the maximum strain energy associated with the motion cycle.

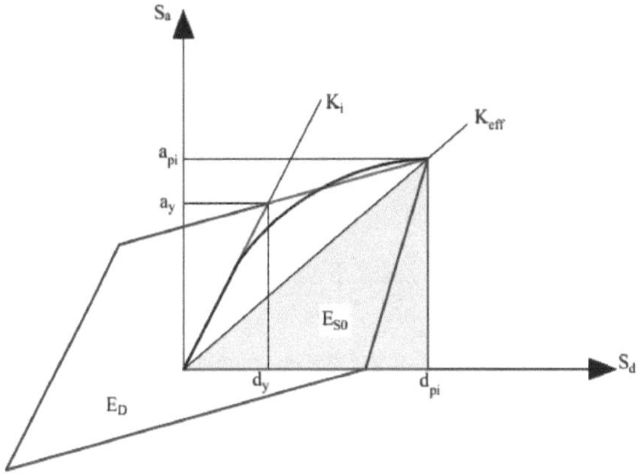

Figure 2.15 Estimation of equivalent viscous damping using the capacitance spectrum method (ATC-40)

- **The performance point of a system with a single equivalent degree of freedom**

The new demand spectrum must intersect the capacity spectrum for the performance point estimate (see Figure 2.16). If the demand spectrum intersects the capacity spectrum within an acceptable tolerance, then the estimate is accepted. If not, the performance point is re-estimated and the procedure is repeated from the step of superimposing the capacity spectrum on the ADRS spectrum.

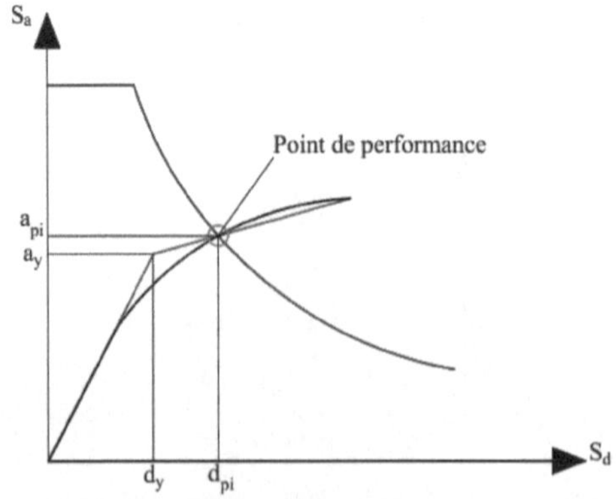

Figure 2.16 Estimation of target displacement (the performance point)

➤ **Performance point of a system with several degrees of freedom**

When the performance point is determined, it is converted to the target displacement of a multi-degree-of-freedom system using the following equation:

$$u_t = PF_1 \emptyset_{ij} S_d \tag{2.41}$$

Where, PF_1 is the participation factor defined in equation (2.37); S_d is the spectral displacement of the equivalent single degree of freedom system, which is defined in equation (2.36).

2.3.4.3. Displacement coefficient method

An effective effort has been made to develop more accurate non-linear procedures by facilitating the idea that the maximum inelastic displacement can be directly estimated by multiplying the corresponding maximum elastic displacement by appropriate modification factors. These factors represent average inelastic-elastic displacement ratios that can depend on natural period (T), displacement ductility ratio (μ) and lateral bending strength. Miranda[144] compared the results obtained using inelastic demand spectra and maximum inelastic-elastic displacement ratios in the capacitance spectrum method, and concluded that the latter approach provides significantly better predictions of the target displacement. This approach is prescribed as the Displacement Coefficient Method in FEMA 273[145] and FEMA 356[86]. As mentioned above, it uses a series of appropriate factors or coefficients (C_0, C_1, C_2 and C_3) to convert the maximum displacement of a linear degree-of-freedom system to a non-linear maximum displacement, which can then be used as the target displacement for the pushover analysis to obtain the overall seismic demands. The modification coefficients also depend on the characteristics of the force-deformation relationship of the idealised non-linear one-degree-of-freedom system. Due to its practical application, this approach has been widely accepted and applied to different types of building structures.

- **FEMA 440 improvements**

FEMA Report 440[146] presented a comprehensive analysis programme to develop improved versions of both the capacitance spectrum method and the displacement coefficient method. This investigation involved determining the maximum non-linear displacements of a large number of one-degree-of-freedom systems subjected to a large number of true ground motion records. To this end, one-degree-of-freedom systems with initial natural periods of between 0.05 and 3.0 seconds, and with nine different levels of lateral resistance, were used. They are modelled with four different hysteretic behaviours (perfectly plastic elastic, stiffness degrading, strength and stiffness degrading, and non-linear elastic behaviour). Based on the results obtained from this comprehensive parametric study, various improvements to both the capacitance spectrum method and the displacement coefficient method have been proposed.

For the capacity spectrum method, a more efficient bilinear approximation of the pushover curve is proposed. Expressions for estimating the effective time period and effective viscous damping (to reduce the elastic spectrum to an inelastic demand spectrum) have also been improved. Similarly, FEMA Report 440[146] also provided improved expressions for displacement modification coefficients for DCM, which were subsequently incorporated into ASCE/SEI 41-06[147] and ASCE/SEI 41-13 .[148]

The coefficient C_3(from FEMA Reports 273[145] and FEMA Report 356[86] , which accounts for increased displacements due to dynamic P-Δ effects) has been suggested to be eliminated and replaced by an upper limit on the normalised failure strength ratio R_{max}(which is intended to account for dynamic instability). If the lateral resistance of a system (normalised by its elastic resistance demand) exceeds R_{max}, a non-linear dynamic procedure should be used to capture resistance degradation and dynamic P-Δ effects to confirm the dynamic stability of the building. The next subsection will discuss the displacement coefficient method as prescribed in ASCE/SEI 41-06[147] and ASCE/SEI 41-13[148] in its most recent form.

- **Non-linear procedures in ASCE/SEI 41-06 and ASCE/SEI 41-13**

The Nonlinear Static Method, defined in ASCE/SEI 41-06[147] , uses the application of a unique invariant loading scheme (based on the first modal form) to the structure, while the target displacement is determined using the improved version of the displacement coefficient method. An inelastic model directly incorporating the non-linear load-deflection behaviours of individual building components is subjected to a monotonically increasing lateral loading pattern (representative of the inertial forces developed during an earthquake), until a target displacement is reached at the control node (typically the roof). This target displacement is intended to represent the maximum displacement likely to be experienced during the design earthquake and is calculated using the expression shown in Figure 2.17.

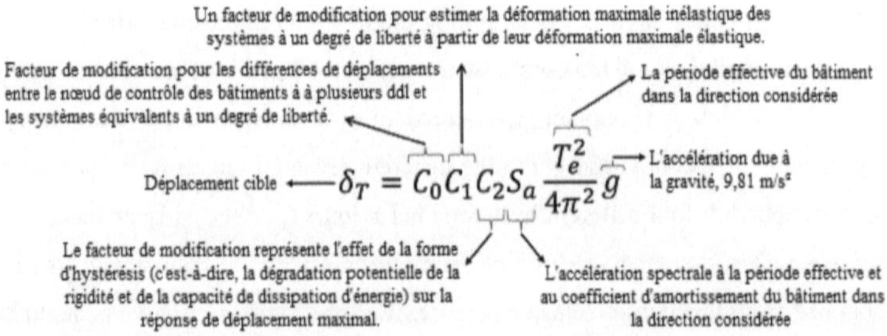

Figure 2.17 Modifying the spectral elastic displacement to estimate the target displacement (Performance point)

According to ASCE/SEI 41-06[147] , it is necessary to carry out the analysis to at least 150% of the target displacement to encourage engineers to investigate the likely performance of the building under extreme loading conditions beyond the design values. Next, the non-linear relationship between the base force and the lateral displacement of the control node is replaced by an idealised force-deflection relationship to determine the effective lateral stiffness (K_e) and the effective yield resistance (V_y) of the building. These parameters are used to determine the effective time period (T_e) and the normalised resistance factor (R). These two parameters (with a site class factor a) are then used to finally determine the displacement modification coefficients $C_1$and C_2. The overall procedure is summarised in Figure 2.18. For periods less than 0.2 seconds, the coefficient $C_1$is taken as the value at T = 0.2 seconds, while for periods greater than 1 second, the equal displacement approximation is assumed to be valid and$C_{(1)}$is taken as 1, indicating that the non-linear peak displacement tends to remain equal to the linear peak displacement. Similarly, for periods greater than 0.7 s, there is no significant hysteretic effect and the coefficient $C_2$is taken as 1. For buildings with negative post-efficiency stiffness, the maximum resistance ratio R_{max}should be determined in accordance with equation (2.42).

$$R_{max} = \frac{\Delta_d}{\Delta_y} + \frac{|\alpha_e|^{-h}}{4} \quad (2.42)$$

Where$\Delta_d$is the lesser of the target displacement (δ_T) or the displacement at maximum base force. WhileΔ y is the displacement at the effective yield strength. h =1+0.15ln(T); andα_e is the post-yield negative effective slope ratio (i.e. the ratio of the post-yield negative slope to the initial line slope in the bilinear approximation).

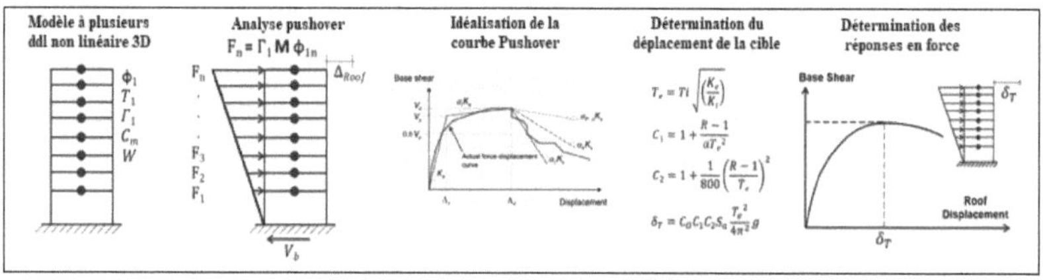

Figure 2.18 An overview of the Nonlinear Static Method (displacement coefficient method) from ASCE/SEI 41-13 [153].

Figure 2.19 shows graphically the expressions from ASCE/SEI 41-13 [153] for the displacement modification coefficients $C_1$and $C_2$for different values of the normalised resistance factor (R) and the site class factor ( ).α

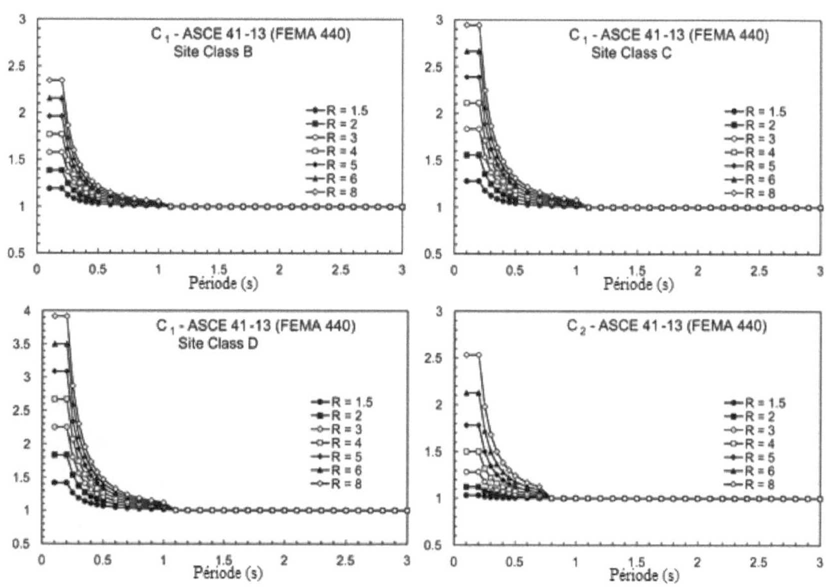

Figure 2.19 The ASCE/SEI 41-13 [153] relationships for the displacement modification coefficients $C_1$and C_2.

2.4. Conclusion

In concluding this chapter, it is clear that seismic analysis is a crucial step in assessing the performance of structures against earthquakes. We have explored various analysis methods, both linear and non-linear, each with its own advantages and limitations. Linear analysis methods, such as linear statics and linear dynamics, provide quick and relatively simple results, but do not always capture the complex behaviour of real structures. In contrast, non-linear analysis methods, such as non-linear statics, allow more accurate modelling of the non-linear behaviours of materials and structural systems. In this chapter, I have placed particular emphasis on the use of non-linear static analysis methods. These methods offer a more realistic representation of the responses of structures to earthquakes, by taking account of geometric and material non-linearities. The non-linear static method used in this book is the N2 method, which was originally proposed by Fajfar, then incorporated into Eurocode 8[133] and recently extended to include frames with infill walls in several studies .[10,141]

Chapter 3: Numerical evaluation of macro-modelling of infill walls incorporating a fibre-discretised plastic ball joint

3.1 Introduction

This chapter is an in-depth exploration of numerical modelling of materials in the context of the structural behaviour of buildings. More specifically, it focuses on the evaluation of macro-modelling of infill walls incorporating a fibre-discretised plastic hinge, a method crucial to the performance of civil engineering structures. By examining the constituent elements of structures, in particular concrete and steel, this study aims to model the non-linear behaviour of the various components accurately and comprehensively, taking into account the cyclic characteristics of materials such as reinforced concrete. The use of the fibre ball model is an essential step in this approach, enabling the non-linear behaviour of structural elements to be derived from the constituent fibres, thereby improving the accuracy of the results.

A review of the existing literature has identified three main numerical methods for modelling infill walls. However, this chapter focuses specifically on the in-depth evaluation of the macro-modelling approach with a discretised fibre section, also known as fibre hinges. Through non-linear analyses performed using the finite element software ETABS, this study seeks to determine the effectiveness of this method in predicting the behaviour of filled porticos. Furthermore, this approach is validated by a previous experimental study conducted by Cavaleri and Di Trapani[149] in 2014. This validation, based on comparisons between the force-displacement curves obtained numerically and the experimental results, reinforces the robustness and reliability of the conclusions drawn from this study.

3.2 Behavioural models

Any structural analysis needs to take into account two fundamental elements:

- The geometry of the structure and the properties of the constituent materials.
- The behaviour of each material is described in the literature by various mathematical laws.

The portal frame structures examined in this book are all made of reinforced concrete. This composite material is made up of two components with distinct characteristics:

- Concrete, whose main function is compression. Its performance in tension is generally very poor, if not negligible. In compression, concrete can be in two distinct states of stress:
  - The first, called "unconfined", corresponds to a situation where the material is not surrounded by longitudinal and transverse reinforcement.
  - The second, called "confined", refers to a situation where the material is encircled by reinforcement.
- Steel, on the other hand, is capable of working in both tension and compression, depending on the stresses involved.

3.2.1. Behaviour of concrete

Many researchers have studied the development of behaviour laws for concrete. For the purposes of this book, we present the Mander model[150] for unconfined and confined concrete.

- **Mander model for the behaviour of unconfined concrete**

The stress-strain diagram for unconfined concrete adopted in our study is that developed by Mander[150] . It consists of a curved section and a linear section, as shown in Figure 3.1.

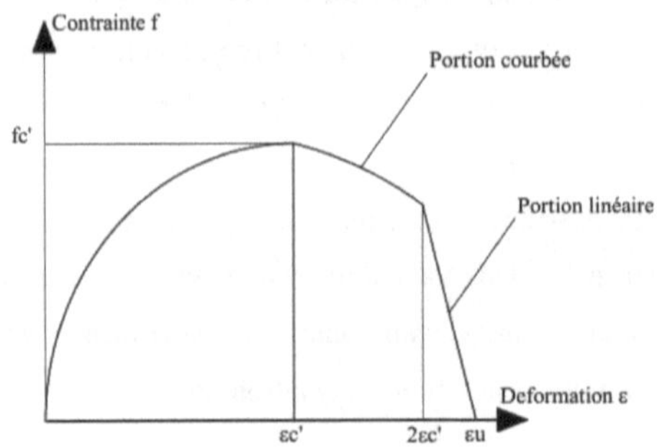

Figure 3.1 Mander model for the behaviour of unconfined concrete [150]

For the stress-strain diagram of unconfined concrete, where$\varepsilon_c \leq 2\varepsilon_c'$, the following equation represents the curved portion of the diagram:

$$f = \frac{f_c' \, x \, r}{r - 1 + x^r} \qquad (3.1)$$

For the stress-strain diagram of unconfined concrete where$2\varepsilon_c' < \varepsilon_c < \varepsilon_u$, the following equation describes the linear part of the graph:

$$f = \left(\frac{2f_c' r}{r - 1 + 2^r}\right)\left(\frac{\varepsilon_u - \varepsilon_c}{\varepsilon_u - 2\varepsilon_c'}\right) \tag{3.2}$$

The variables x and r are given by the following equations:

$$x = \frac{\varepsilon_c}{\varepsilon_c'} \tag{3.3}$$

$$r = \frac{E_c}{E_c - \frac{f_c'}{\varepsilon_c'}} \tag{3.4}$$

Where:

f Compressive strength of concrete.

ε_c Compressive deformation of concrete.

f_c' Compressive strength of concrete.

ε_c' Compressive deformation of concrete corresponding to .f_c'

ε_u Ultimate compressive strain capacity of concrete.

E_c Modulus of elasticity of the initial concrete.

- **Mander model for the behaviour of confined concrete**

Determining the stress-strain relationship of confined concrete poses the main difficulty of determining its compressive strength under confinement, noted f_{cc}' . This data can be obtained in various ways. As early as 1928, Richart[151] proposed a formulation in which certain parameters were adjusted experimentally. Based on this earlier work, Mander[150] developed a method for estimating f_{cc}' . The following is a presentation of Mander's model for confined concrete.

The fundamental equation under monotonic compression represents a unified stress-strain approach for confined concrete with rectangular-shaped transverse reinforcement, as suggested by Mander. Based on the equation proposed by Popovics[152] , the stress-strain model proposed by Mander for both confined and unconfined concrete under monotonic loading is shown in Figure 3.2.

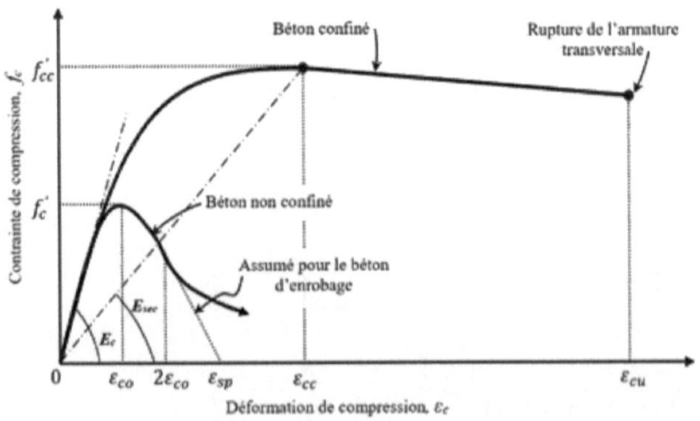

Figure 3.2 Mander model for the behaviour of confined and unconfined concrete [150]

The longitudinal compressive stress in concretef_c , for a quasi-static strain rate and monotonic loading, is given as a function of the compressive strainε_c , by the following formula:

$$f_c = \frac{f'_{cc}\, x\, r}{r - 1 + x^r} \tag{3.5}$$

The variable$x$ is given as follows:

$$x = \frac{\varepsilon_c}{\varepsilon_{cc}} \tag{3.6}$$

Where:

f'_{cc} Compressive strength of confined concrete.

ε_c Longitudinal compression deformation of concrete.

ε_{cc} The deformation corresponding to the maximum stress in the concrete is given below:

$$\varepsilon_{cc} = \varepsilon_{c0}\left[1 + 5\left(\frac{f'_{cc}}{f'_{c0}} - 1\right)\right] \tag{3.7}$$

f'_{c0} Strength of unconfined concrete.

ε_{c0} Deformation of unconfined concrete.

The variable$r$ is given by the following relationship :

$$r = \frac{E_c}{E_c - E_{sec}} \tag{3.8}$$

E_{sec} The secant modulus of the confined concrete corresponding to the peak stress is given by the following relationship :

$$E_{sec} = \frac{f'_{cc}}{\varepsilon_{cc}} \tag{3.9}$$

For the stress-strain diagram for unconfined concrete, where$\varepsilon_c > 2\varepsilon_{c0}$, the behaviour of the embedding concrete is assumed to be linear, reaching the zero stress point for a strain (ε_{sp}) corresponding to the bursting of the compressed concrete.

The effective lateral confinement pressure and the confinement efficiency coefficient refer to the way in which the maximum pressure of the transverse confinement reinforcement is exerted on a specific surface in the plane of the section. This surface is delimited by parabolic arcs connected at the longitudinal reinforcement. The confinement stress is fully developed due to the bending action. The same phenomenon is observed between the transverse reinforcements in the longitudinal direction of the part, as illustrated in Figure 3.3.

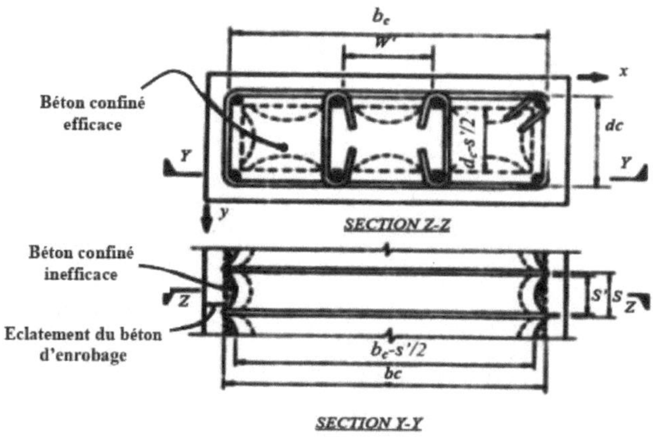

Figure 3.3 Effective confinement concrete for rectangular transverse reinforcement .[150]

Using the stress-strain relationship given by formula (3.5) to determine the ductility and strength of columns, we assume that the confined concrete cross-section, A_{cc}, is that bounded by the transverse reinforcement. The effective lateral confining pressure is then given by the following formula when $A_{cc} > A_e$:

$$f_l' = f_l \; k_e \tag{3.10}$$

Where:

f_l Lateral pressure of the transverse steels, assumed to be uniformly distributed around the surface of the confined concrete.

k_e The containment efficiency coefficient is given as follows:

$$k_e = \frac{A_e}{A_{cc}} \tag{3.11}$$

A_e Effective confined concrete surface.

A_{cc} The section of confined concrete delimited by the transverse reinforcement is given by the following formula as follows :

$$A_{cc} = A_c(1 - \rho_{cc}) \tag{3.12}$$

A_c Surface of the concrete confined between the axes of the transverse reinforcement.

ρ_{cc} Ratio of the cross-sectional area of longitudinal reinforcement to the surface area of confined concrete.

The compressive strength of the confined concretef'_{cc} , is determined as a function of the effective lateral confining stressf'_l . Generally, the compressive strength of the confined concrete is deduced using the following relationship :

$$f'_{cc} = f'_{c0}\left(2.254\sqrt{1 + \frac{7.94\,f'_l}{f'_{c0}}} - 2\frac{f'_l}{f'_{c0}} - 1.254\right) \tag{3.13}$$

Where:

f'_{c0} Strength of unconfined concrete.

3.2.2. Behaviour of steel

The diagram in Figure 3.4 illustrates the behaviour of steel subjected to a tensile test. This diagram is generally divided into two or three distinct parts:

1) The linear elastic part, characterised by the Young's modulus of elasticity (E) and by the yield point, also known as the elastic limit (f_y).

2) The almost horizontal plastic part, which is at a stress level equivalent to that of the elastic limit. It should be noted that not all steels have this feature.

3) The strengthening (or strain-hardening) part, characterised by an increase in stress up to the maximum. When the stress reaches its maximum, the strain appears, accompanied by a decrease in stress until fracture.

In this context, we present below the Park[153] model for steel used in this study.

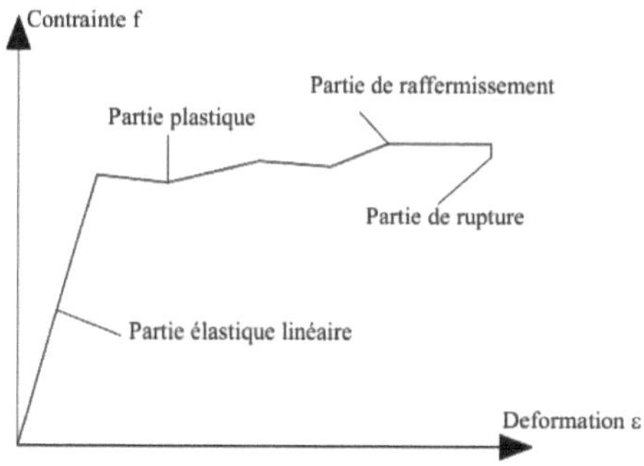

Figure 3.4 Stress-strain diagram for steel in tension.

- **Park's model for the behaviour of steel**

The Park model[153] for steel is used in this study. Its stress-strain behaviour law is shown in Figure 3.5. This Park's law[153] is used for both tensile and compressive steel.

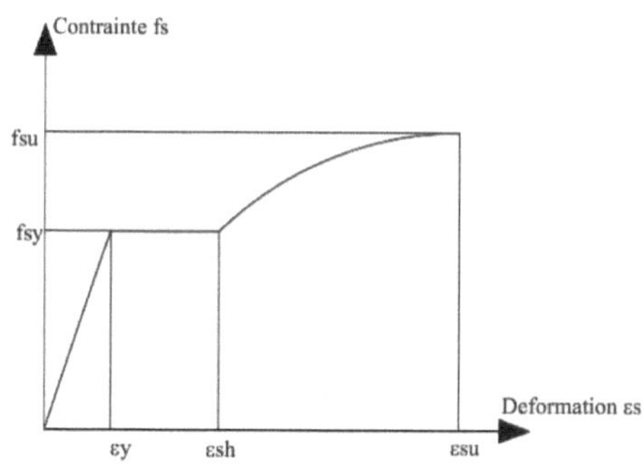

Figure 3.5 Park's model for steel behaviour [153]

Where:

ε_yLimited elastic deformation of steel.

ε_{sh} Deformation of the steel corresponding to the start of work hardening.

ε_{su} Ultimate deformation of steel.

f_{sy}Elastic limit stress of the steel.

f_{su} Ultimate stress in steel.

E_s Modulus of elasticity of steel.

The equations defining the stress-strain relationships of the behaviour law for steel are as follows:

For $\varepsilon_s \leq \varepsilon_y$

$$f_s = E_s \varepsilon_s \tag{3.14}$$

For $\varepsilon_y < \varepsilon_s \leq \varepsilon_{sh}$

$$f_s = f_{sy} \tag{3.15}$$

For $\varepsilon_{sh} < \varepsilon_s \leq \varepsilon_{su}$

$$f_s = f_{sy}\left[\left(\frac{m(\varepsilon_s - \varepsilon_{sh}) + 2}{60(\varepsilon_s - \varepsilon_{sh}) + 2}\right) + \left(\frac{(\varepsilon_s - \varepsilon_{sh})(60 - m)}{2(30r + 1)^2}\right)\right] \tag{3.16}$$

Where:

$$m = \frac{\left(\frac{f_{su}}{f_{sy}}\right)(30r + 1)^2 - 60r - 1}{15r^2} \tag{3.17}$$

$$r = \varepsilon_{su} - \varepsilon_{sh} \tag{3.18}$$

3.3 Non-linear modelling of filled portals

This section describes the procedure used to model the filled portal frames. The assumptions underlying the modelling approach are as follows: columns, reinforced concrete beams and masonry infills were modelled using a distributed plasticity model (fibre hinges) to describe their non-linear behaviour. The columns and beams were represented as reinforced concrete elements, while the masonry infills were represented as diagonal compression rods centred on the columns. This modelling approach is illustrated in Figure 3.6.

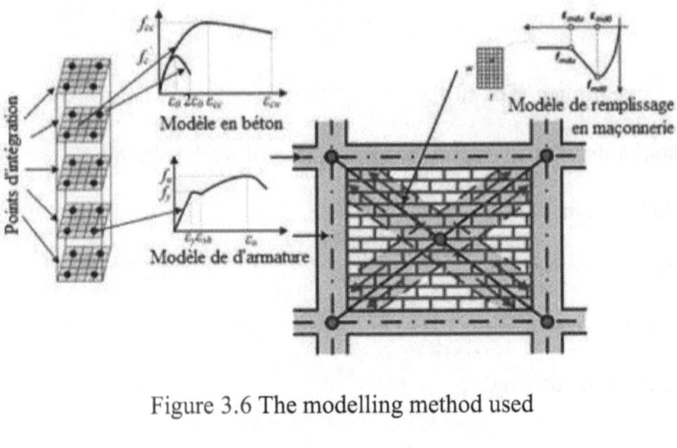

Figure 3.6 The modelling method used

In order to determine the stress-strain relationship to be assigned to the equivalent diagonal connecting rod fibre sections, the model proposed by Di Trapani et al.,[119] was adopted in this study (see Figure 3.7).

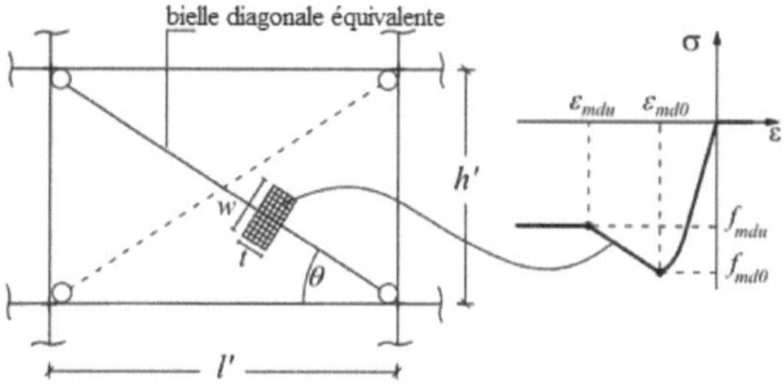

Figure 3.7 The model proposed by Di Trapani et al.[119] for an equivalent diagonal connecting rod

The model is defined by the maximum strength (f_{md0}), maximum deformation (ε_{md0}), ultimate strength (f_{mdu}) and ultimate deformation ($\varepsilon_{(mdu)}$) of the material, which are obtained using the following semi-empirical equations:

$$f_{md0} = 26.9\,\tilde{f}_m.\alpha^{-0.287} \tag{3.19}$$

$$f_{mdu} = f_{md0}(0.043\beta - 0.06) \tag{3.20}$$

$$\varepsilon_{md0} = 3.024.\varepsilon_{m0}.\gamma^{0.347} \tag{3.21}$$

$$\varepsilon_{mdu} = 0.0184.\varepsilon_{md0}.\delta^{-1.166} \tag{3.22}$$

The above parameters are determined directly from the mechanical and geometric properties of the filled portal using the following equations:

$$\alpha = \frac{\tilde{f}_m^2.w.t}{(f_{vm} + \mu.\sigma_n)^{0.2}(l/h).\lambda^{*0.2}} \tag{3.23}$$

$$\beta = \frac{f_{md0}^{0.7}.w.t}{\tilde{E}_m^{0.2}.d} \tag{3.24}$$

$$\gamma = \left(\frac{f_{mdu}^2}{f_{md0}}\right)\left(\frac{E_c}{\tilde{E}_m^{1.5}}\right) \tag{3.25}$$

$$\delta = \tilde{E}_m^{0.2}.\varepsilon_{md0} \tag{3.26}$$

The equations used to determine the parameters include several variables. One of these variables is $\varepsilon_{(md0)}$ which represents the maximum deformation of the masonry in compression and is generally assumed to be equal to 0.0015. Add to this $\tilde{E}m$ and $\tilde{f}m$ which represent the modulus of elasticity of the masonry and the compressive strength, respectively. These values are used to summarise the stiffness and compressive strength of masonry in different orthogonal directions, as expressed in the following equations:

$$\tilde{E}_m = \sqrt{E_{m1} \cdot E_{m2}} \tag{3.27}$$

$$\tilde{f}_m = \sqrt{f_{m1} \cdot f_{m2}} \tag{3.28}$$

The modulus of elasticity and strength of the masonry along the two orthogonal directions are represented by E_{m1}, E_{m2}, f_{m1} and f_{m2}. The cross-sectional geometry of the equivalent diagonal was identified using the method suggested by Asteris et al.[154] . The thickness of the diagonal (t) is equal to the thickness of the infill wall, while the width (w) is determined using the semi-empirical equation Eq. 3.29 :

$$w = \kappa^{\gamma^*} \left(\frac{h}{l}\right) \frac{c^*}{\lambda^{*\beta^*}} d \tag{3.29}$$

The equation for determining the width (w) of the equivalent diagonal cross-section comprises several variables. These variables included d , which represents the length of the equivalent diagonal cross-section; h and l , which represent the actual length and height of the masonry wall respectively. In addition, the equation involves a coefficient κ , which is used to take into account the lateral stiffening effect exerted by the vertical load. The value of κ is estimated using the following expression:

$$\kappa = [1 + (18\lambda^* + 200)\varepsilon_v] \tag{3.30}$$

Consequently, the axial deformation ε_v is calculated by dividing the total load F_v acting on the two boundary columns by their cross-sectional area A_c and their modulus of elasticity E_c . Thus, ε_v can be expressed as follows:

$$\varepsilon_v = \frac{F_v}{2E_c A_c} \tag{3.31}$$

The formula suggested by Papia et al.[155] is used to calculate the parameter λ^*, which is expressed as follows:

$$\lambda^* = \frac{\tilde{E}_m}{E_c} \frac{th'}{A_c} \left(\frac{h'^2}{l'^2} + \frac{1}{4} \times \frac{A_c}{A_b} \times \frac{l'}{h'}\right) \tag{3.32}$$

Where: (h') is the actual length of the columns; (l') is the actual length of the beams.

The variablesA_c andA_b represent the cross-sectional areas of the columns and beam respectively. The exponentγ * in equation 3.29, is used to take into account the combined effect of the openings and the aspect ratio of the infill, and is determined by equation 3.33.

$$\gamma^* = 1 + 0.5\frac{r}{(l/h)^4} \tag{3.33}$$

In the real case of solid fillings, r is equal to 1. The parameters$c$ * andβ * used in the expression of (w) are estimated using the following equations:

$$c^* = 0.249 - 0.0116v + 0.567v^2 \tag{3.34}$$

$$\beta^* = 0.146 - 0.0073v + 0.126v^2 \tag{3.35}$$

The calculation of the parameters c* andβ * depends on the Poisson's ratio (ν) of the infill walls, which is evaluated along the diagonal direction of the infill. In this study, for simplicity, the average Poisson's ratio along the two orthogonal directions is used. If Poisson's ratio data is not available, a value of 0.1 is used as an approximation.

3.4. Experimental part

Three in-plane experimental tests on filled portals were selected as a reference. The experimental tests were chosen from the experimental campaigns carried out by Cavaleri and Di Trapani[149] . The specimens were selected with the aim of covering as much as possible the different types of masonry in the masonry infills. Specimens S1A, S1B and S1C from Cavaleri and Di Trapani[149] were arranged with limestone, hollow clay and lightweight concrete masonry units respectively. In addition, general details of the specimens are provided in Figure 3.8. The mechanical properties of the materials are provided in Table 3.1, and for further details readers may refer to the original article.

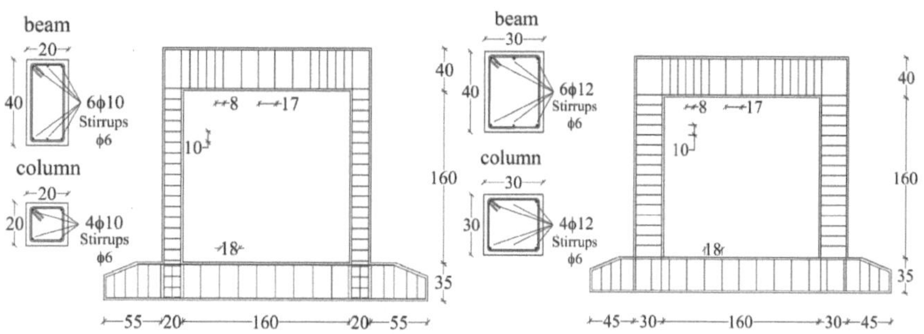

(a) S1A and S1B (b) S1C

Figure 3.8 General details of the specimen (all dimensions in Centimetres) [149]

Table 3.1 Mechanical properties of materials for specimens [149]

		Specimens		
		S1A	**S1B**	**S1C**
Masonry	**Type**	*Limestone*	*Hollow clay*	*Light concrete*
	t (mm)	210	150	250
	$E_{(1)}$ (MPa)	7408	5038	1944
	$E_{(2)}$ (MPa)	3933	6401	4565
	$f_{(m)\,(1)}$ (MPa)	3.08	4.18	0.30
	$f_{(m)\,(2)}$ (MPa)	2.67	8.66	1.74
	$f_{(vm)}$ (MPa)	0.73	1.07	0.29
	v	0.15	0.10	0.17
Concrete	**$f_{(c)}$ (MPa)**	25	25	25
	$E_{(c)}$ (MPa)	25500	25500	25500
Armature	**$f_{(y)}$ (MPa)**	450	450	450
	$E_{(s)}$ (MPa)	210000	210000	210000

3.5. Model validation: experimental/analytical comparisons

The analysis of the experimental test model was carried out in two stages. Firstly, vertical loads were applied to the top of the columns (200 kN on each column) to reproduce gravity loading conditions. In a second step, a monotonically increasing horizontal displacement was assigned. In Figure 3.9, the force-displacement curves obtained from the analysis are compared with the experimental positive and negative monotonic envelopes. As can be seen, the numerical responses adequately reproduce the average experimental trend in terms of maximum strength and stiffness.

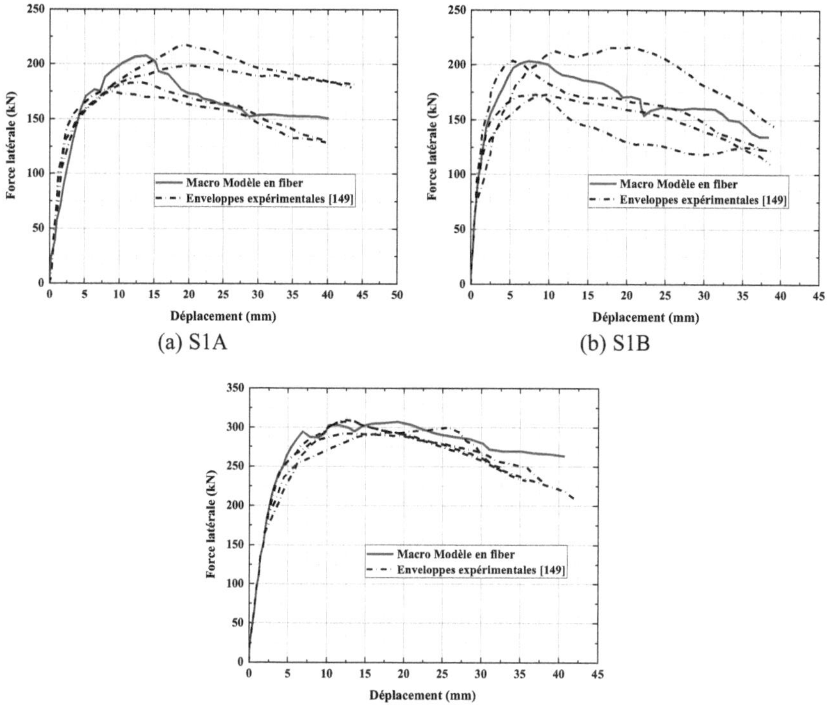

(a) S1A

(b) S1B

(c) S1C

Figure 3.9 Experimental/digital comparisons

3.6. Conclusion

In this chapter, the focus has been on assessing the effectiveness of numerical modelling using the macro-modelling approach with a discretised fibre section (fibre hubs) for predicting the behaviour of filled porticos. Non-linear analyses were carried out using ETABS finite element software, and this approach was validated by comparing it with an experimental study conducted by Cavaleri and Di Trapani[149] based on force-displacement curves. This validation confirms the model's ability to reliably reproduce the behaviour of filled portal frames, paving the way for further exploration in the next chapter to understand the effects of the arrangement of infill walls in plan and/or elevation on the seismic performance of reinforced concrete structures.

Chapter 4: Assessment of the effects of the layout of infill walls in plan and/or elevation on the seismic performance of reinforced concrete structures

4.1 Introduction

This chapter evaluates the effects of infill wall arrangement in plan and/or elevation on the seismic performance of a reinforced concrete structure located in a high seismicity zone. A pushover analysis procedure is carried out to evaluate the seismic performance of the reinforced concrete structure with different infill wall arrangements, and to study the results related to capacity curves, absorbed energy, inter-storey displacement and performance levels. Then it is also necessary to check their effect on torsion in the building. Therefore, a four-storey reinforced concrete building with masonry walls was analysed, taking into account their distribution on the sides of the building in a symmetrical and asymmetrical way, and with variable availability in different floors and bays.

4.2. Description of the building structure

A four-storey building of the same height (h = 3 m) with five bays of equal length in both directions, as shown in Figure 4.1, was studied. Thirteen models of reinforced concrete structures were developed in the portal frame of this study, taking into account the arrangement of the twenty-five centimetre thick infill walls on the sides in the X direction of the building (***first and last axes, i.e. axes 1 and 6 as shown on the plan in Figure 4.1***) in a symmetrical and asymmetrical manner, with variable availability in the different storeys and bays (see Figure 4.2). With: (M0) building model without masonry infill walls; (from M1 to M12) building models with infill wall layout in plan and/or elevation.

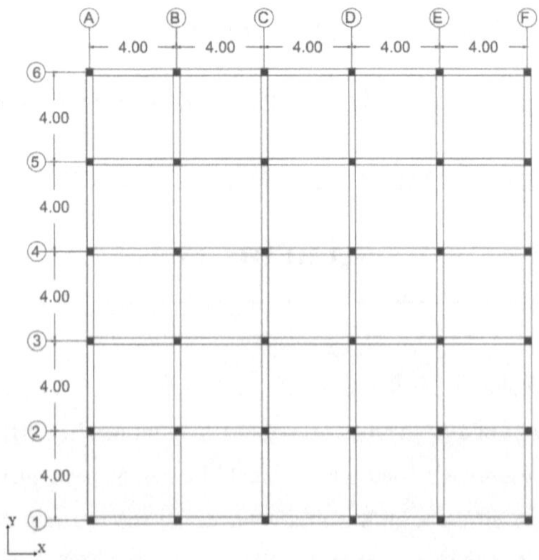

Figure 4.1 Schematic view of the respective structure (in m).

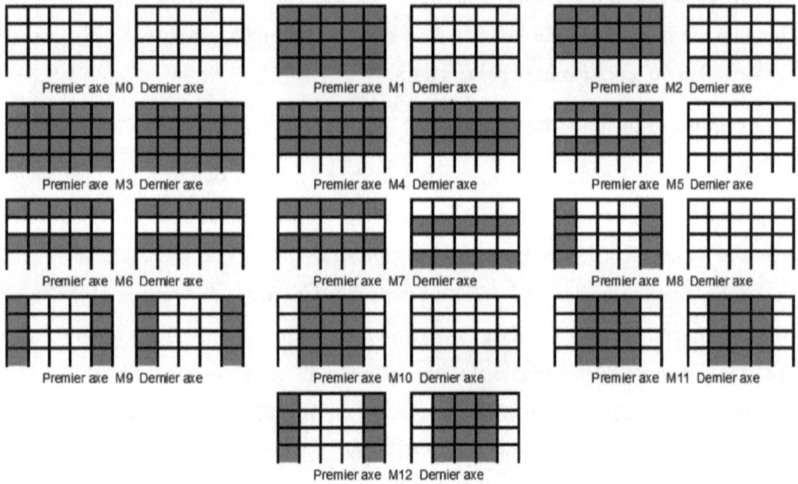

Figure 4.2 Layout of the infill walls in the models.

Modelling was carried out using ETABS software. The design and analysis of the models were carried out in accordance with the Algerian Paraseismic Rules (RPA2003)[156] and the calculation rules, known as the BAEL 91 Rules[157] . The building is located in a high seismicity zone (Zone III design spectrum according to RPA2003), as shown in Figure 4.3, so the building has residential use (Group 2), 5% viscous damping and loose soil (S3). At the analysis stage of the proposed models, a permanent load of 6 kN/m² and an operating load of 1.50 kN/m² were selected. The vibrating mass is calculated using the permanent load and 20% of the operating load, as specified in RPA2003[156] . The mechanical properties of the materials used in the structural elements are given in Table 4.1, and the properties of the clay brick masonry used to construct the infill walls are presented in Table 4.2, while Table 4.3 presents the cross-sectional data for the structure.

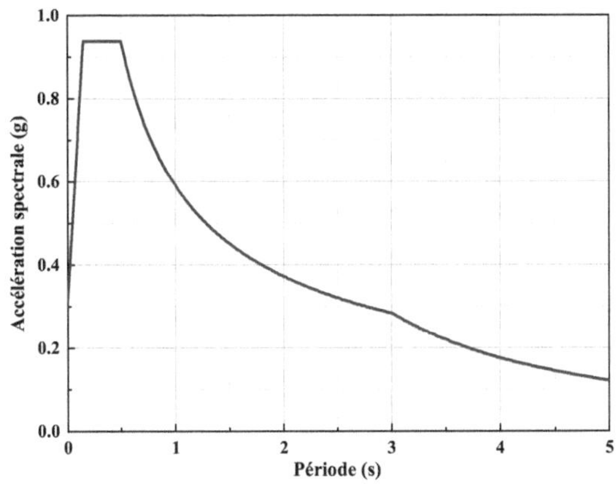

Figure 4.3 Zone III design spectrum based on RPA2003 .[156]

Table 4.1 Mechanical properties of materials.

Concrete		Reinforcing bar	
$F_{(c)}$ (MPa)	$E_{(c)}$ (GPa)	$F_{(y)}$ (MPa)	$E_{(s)}$ (GPa)
25	32	400	210

Table 4.2 Properties of clay brick masonry .[149]

$f_{(m)(1)}$ (MPa)	$f_{(m)(2)}$ (MPa)	$E_{(m1)}$ (MPa)	$E_{(m2)}$ (MPa)	$f_{(mv)}$ (MPa)	N
4.18	8.66	5038	6401	1.07	0.10

Table 4.3 Structural cross-section data.

	Section (cm²)	Reinforcement		
		Longitudinal		Transversal
Posts	40 × 40	8∅ 14		∅ 8
Beams	30 × 40	Sup: 3∅ 12	Inf: 3∅ 12	∅ 8

4.3. Description of the numerical model

The assumptions underlying the modelling approach are as follows: columns, reinforced concrete beams and masonry infill walls were modelled using a distributed plasticity model (fibre hinges) to describe their non-linear behaviour. The columns and beams were represented as portal frame elements, while the masonry infill wall was represented as equivalent diagonal rods centred on the columns. This modelling approach was detailed and presented in the previous chapter.

The cross-section of the portal members and masonry infill wall was divided into fibres, each with a unique material property appropriate for reinforcing steel, masonry infill wall, or unconfined or confined concrete. The constitutive law for the portal frame elements was based on the work of Mander et al.[150] , while the constitutive law for the masonry infill wall was defined by Di Trapani et al.[119] . The validity of this model is limited to the in-plane response of portal frames with infill to provide the assessment of 3D portal frames with infill subjected to earthquakes. The cross-section of each column was divided into 8 fibres for unconfined concrete, 9 fibres for confined concrete, and a number of reinforcement fibres depending on the number of longitudinal reinforcement bars. The cross-section of each beam was divided into 8 fibres for unconfined concrete, 9 fibres for confined concrete, and 4 reinforcement fibres. The equivalent diagonal beams were divided into 21 fibres.

4.4 Methodology

4.4.1. Non-linear static analysis

Most engineers consider the use of non-linear static analysis methods in structural analysis to be important and unavoidable in order to reduce the problems of linear dynamic analysis. Compared with traditional static analysis, this type of study makes it easier to interpret the results and provides more details about the seismic behaviour of a structure. The most accurate seismic analysis method for structures is non-linear dynamic analysis, which uses seismic recordings to obtain the necessary assessment parameters. However, it can be time-consuming and economically unjustifiable, particularly for structures with 3D models[126] . In addition, to increase accuracy, several seismic recordings with similar construction site conditions must be used, further complicating the analysis process.

For the first time after 1970, a non-linear static analysis (Pushover) was put into practice. This is a method that stresses structural models under additional lateral loads to achieve a target displacement or structural instability. The aim is to make the target displacement equal to the maximum displacement of an expected earthquake. Rooshenas[126] has reported that the development of non-linear computer software and the concept of performance-based engineering have led to an increasing use of non-linear analyses, such as Pushover analysis. In this paper, a static Pushover analysis was used. It is a method for assessing the seismic performance of buildings and for designing new structures[158] , and has been introduced in many recent seismic regulations and guidelines[86,132,133] . The non-linear static method used in this paper is the N2 method which was originally proposed by Fajfar and Gašperšič[159] , then incorporated into Eurocode 8[133] and recently extended to encompass portal frames with infill walls in several studies . [10,141]

The 3D structure is loaded in an additional acceleration loading mode according to its fundamental mode in the "X" axis direction controlled by the target displacement, including the effect of P-Δ. Then, use the N2 method to determine the capacity curve with a stop at the performance point that defines both the maximum displacement and the maximum base force of the structure. Then the energy absorption can be defined as the area under the capacity curve. In addition, it is common practice in seismic engineering to indicate damage states in a structure by performance levels as recommended by FEMA Guidelines 356[86] for the building system. In addition, the performance levels take into account only the structural elements, without including the infill walls.

4.4.2. Checking torsion in the structure

The seismic response of buildings can be significantly altered by torsional effects, which have been responsible for serious damage and even the complete collapse of structures in previous earthquakes. These effects can arise from a variety of causes, including non-uniform distribution of mass, stiffness and strength in the structure. It is emphasised that the importance of torsional effects was demonstrated by the massive damage caused during the Michoacán earthquake in Mexico in 1985; this event highlighted the need to better understand the problem and improve design requirements[160] . Torsional effects generally decrease as plastic deformation increases. To account for the influence of torsion, a linear modal spectral analysis of the 3D model is required to determine torsional amplifications. These amplifications represent an upper limit of the torsional amplifications in the majority of cases[161] . A linear modal spectral analysis is performed independently in this study for excitations in the horizontal X direction, where the results are combined according to the SRSS rule. For ease of presentation and discussion of the results, we rely on the torsional amplification factor relationship (A_x)

described in ASCE 7-10[162] to determine the torsional coefficient (C_T). This coefficient, given by equation 1, is intended to verify whether or not torsion is present.

$$C_T = \frac{\delta_{max}}{\delta_{avg}} \tag{4.1}$$

Where: δ_{max} is the maximum displacement at level (mm); δ_{avg} is the average displacement at the extreme points of the floor of the structure at level (mm).

When the torsion coefficient (C_T) exceeds 1.20, this means that the model analysed is subject to torsion.

4.5. Results and interpretation

4.5.1. Non-linear static analysis

The results obtained from the non-linear static analysis carried out to determine the behaviour of 13 models in the X direction will be discussed, and then the performance point of each model will be determined using the N2 method to evaluate their performance. Given the similarities in the results obtained from this analysis and in order to clearly differentiate between the models, the models have been grouped according to the arrangement of the infill walls in three study cases as shown in Table 4.6. The results obtained are capacity curves showing the lateral stiffness, maximum displacement and maximum base force of the building, as well as the absorbed energy, inter-storey displacement and performance levels to determine the damage states of the building. Performance levels are categorised into four levels according to FEMA 356[86] : operational (A-B), immediate occupancy (IO), occupant safety (LS) and collapse prevention (CP). In addition, the performance levels take into account only the structural elements, without including the infill walls.

Table 4.6 Case studies.

1st case study	*2nd case study*	*3rd case study*
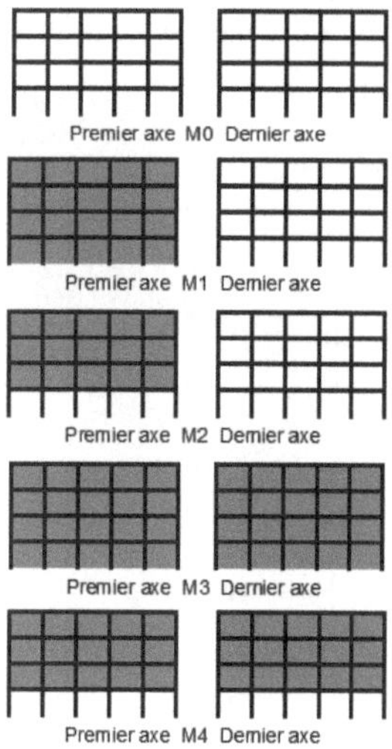	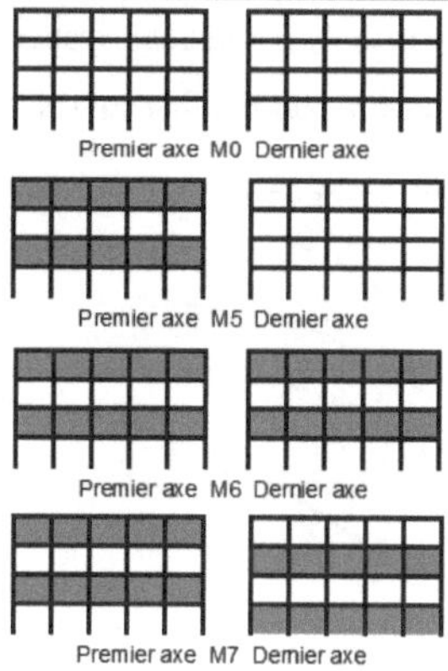	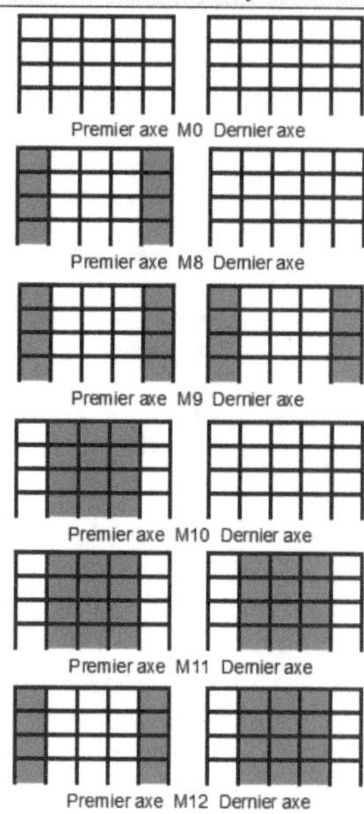

4.5.1.1. First case study

The capacity curves are clearly shown in Figure 4.4. From these curves, it can be seen that the infill walls resulted in a significant increase in the lateral stiffness of the building. Such results are in agreement with the published literature[1–4,10,47,154,155] . Models M1 and M3 showed the largest increase in lateral stiffness of 336 and 387%, respectively, while models M2 and M4 showed a smaller increase of 156 and 170%, respectively. As a result, the maximum building displacement decreased by varying percentages proportional to the availability of the infill walls. However, there is a loss of initial stiffness between models M1 and M2, and also between models M3 and M4, of 41 and 45%, respectively, indicating that the lack of infill walls at ground floor level reduces lateral stiffness. This result was confirmed for 2D frames by Adibi et al. . [120]

When compared to the maximum base shear force, a marked increase in the maximum base shear force was recorded for models M1 and M3 of 19 and 34%, respectively, compared to model M0, with a slight increase of 2 and 3% in models M2 and M4. This indicates that the presence of infill walls has an effect on the determination of the true base shear value of the building, particularly when infill walls are present at ground floor level. It can also be seen that there is a decrease in the maximum displacement value of 49% and 39% for models M3 and M4, respectively. Compared to model M0, models M1 and M2 show a slight decrease in maximum displacement of approximately 5 and 15%, respectively, since there is no masonry infill on the sides of the structure and in all the filled compartments.

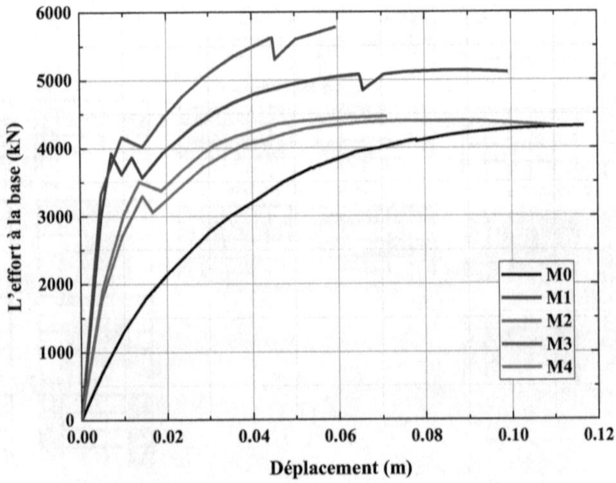

Figure 4.4 Capacity curves

Compared with the energy absorbed in model M0 (see Figure 4.5), a relative increase of 17 and 10% was recorded for models M1 and M2, respectively; on the other hand, a relative decrease of 28 and 31% was recorded in models M3 and M4, respectively. This is due to the fact that the M1 and M2 models with an asymmetrical layout in the plane have a large displacement compared to the models with a symmetrical layout in the plane.

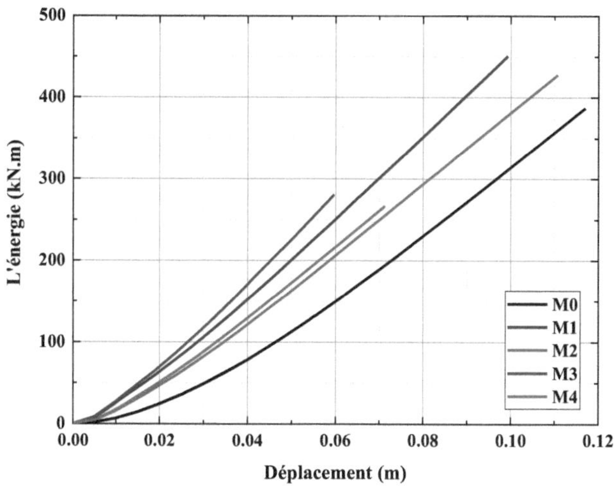

Figure 4.5 Absorbed energy

Figure 4.6 shows the maximum inter-storey displacement (DIEmax) of the models. For models M1 and M2, the DIEmax increased by 50 and 42%, respectively, compared to model M0, while it decreased by 48 and 11% for models M3 and M4, as they are symmetrical in the plane. This indicates that there is a proportionality between the energy absorbed by the building and the DIEmax, which can be explained by the availability and distribution of the infill walls in the building.

Models M0, M1 and M4 achieved LS performance levels, while model M3 achieved an IO performance level, and model M2 was the only one in which one of its elements achieved a CP performance level (see Figure 4.7). Therefore, infill walls can have a positive effect on structural response, provided that they are regularly arranged in the structure. This result is consistent with the result of the study[10] for the case of 2D frames. The lack of infill walls at ground floor level also leads to a decrease in performance between models M1 and M2, and a decrease between models M3 and M4 for the same reason. This result is also consistent with that reported by Adibi et al.[120] for 2D frames.

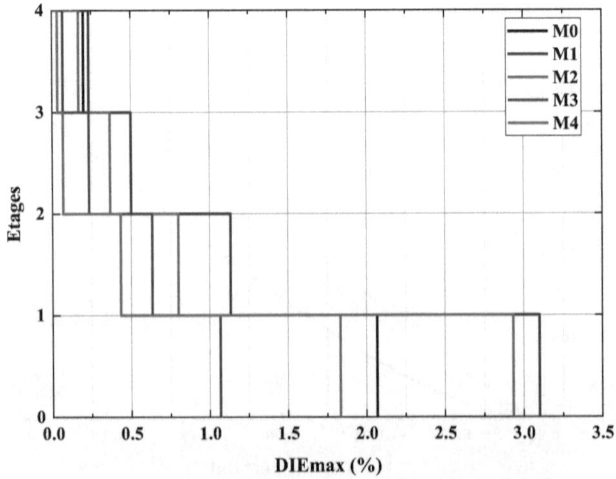

Figure 4.6 Maximum inter-storey displacement

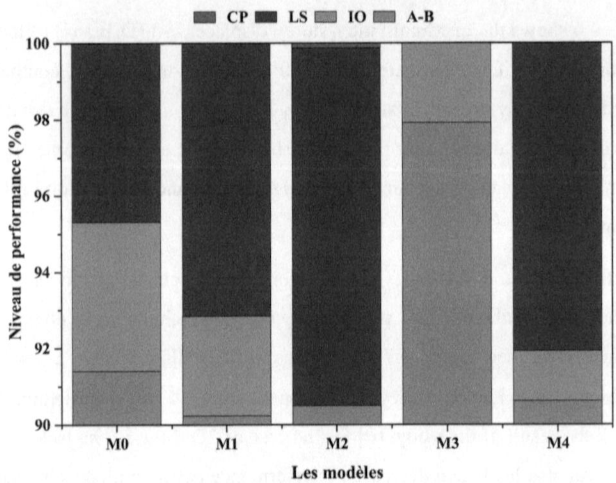

Figure 4.7 Performance levels

In addition, the damage states for these models in the first and last axes that contain the layout of the infill walls in plan and/or elevation are given in Table 4.7.

Table 4.7 Damage states

Models	First axis (Axis 1)	Last axis (Axis 6)
M0	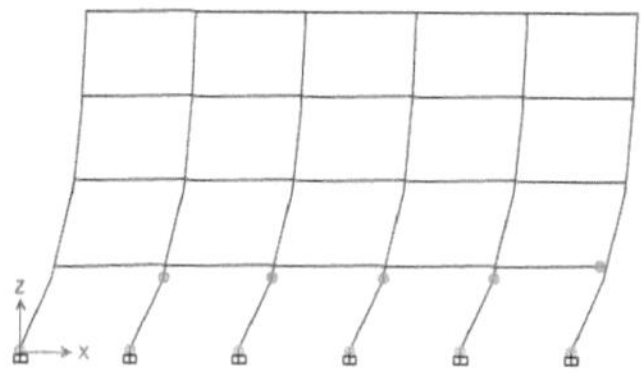	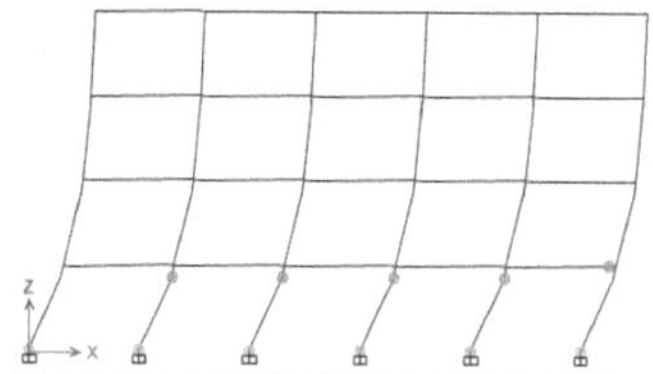
M1	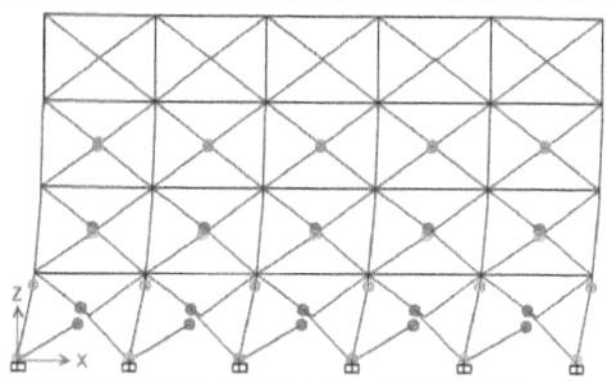	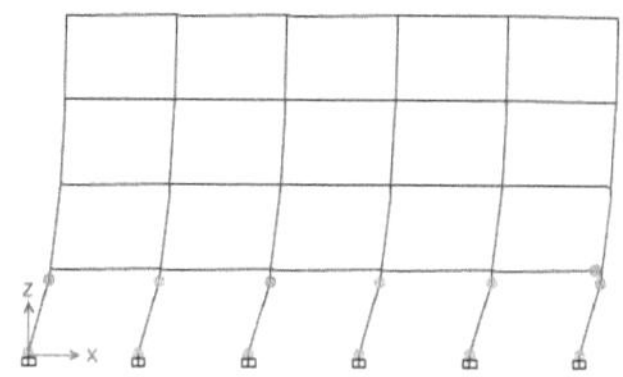
M2	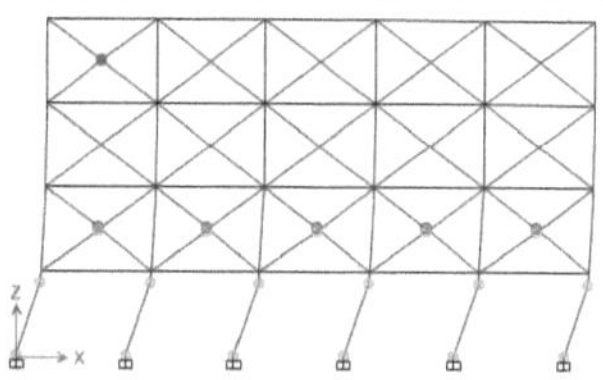	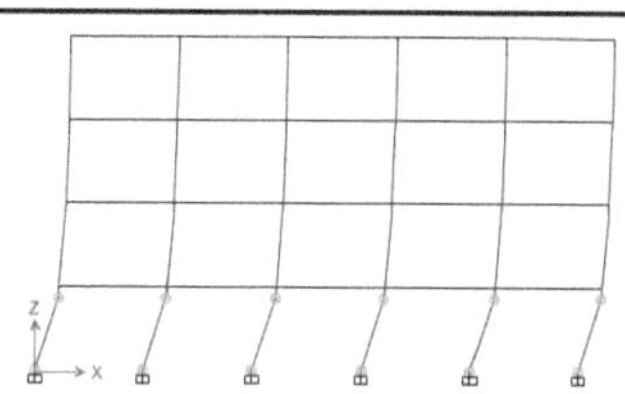
M3	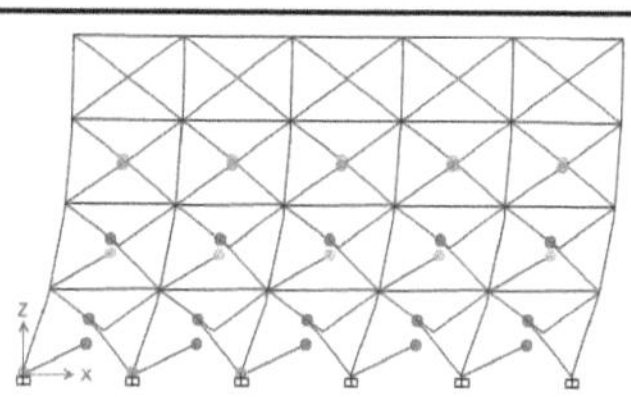	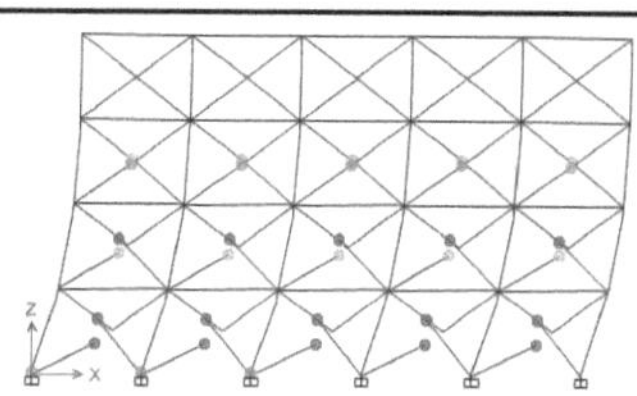
M4	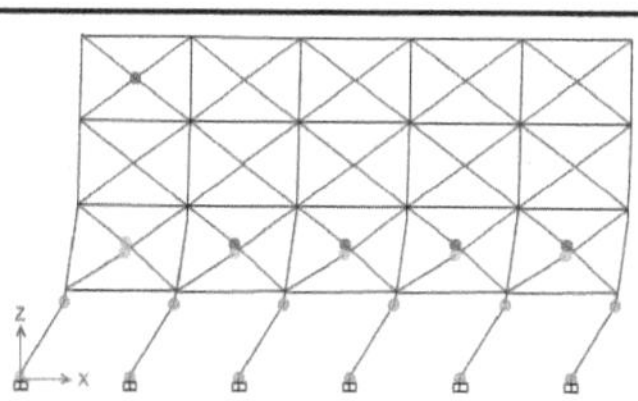	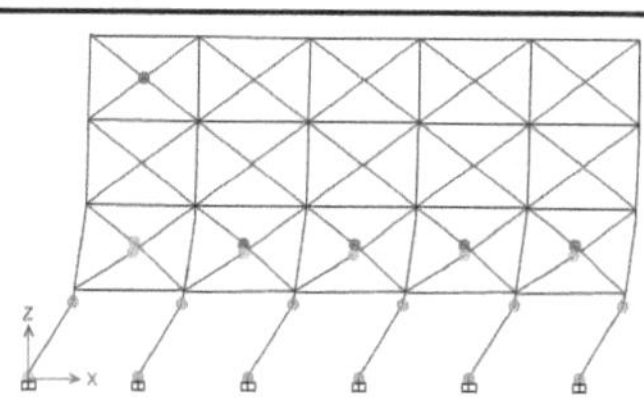

4.5.1.2. Second case study

The capacity curves for the second study case are shown in Figure 4.8. From the results, it can be seen that there is an increase in initial lateral stiffness for models M5, M6 and M7 of approximately twice that of model M0. In addition, there is a decrease in maximum displacement of 30 and 28% for models M6 and M7, respectively, compared with model M0.

Model M5 shows a negligible decrease of 4%, due to the number of infill walls, which are only available in the first frame of the building, unlike models M6 and M7, which have infill walls in the first and last frames of the building. There is also a slight difference in maximum displacement between models M6 and M7, as the first and last frames have the same number of infill walls with an asymmetric distribution. This is because the infill walls of a building increase its rigidity and reduce its displacement.

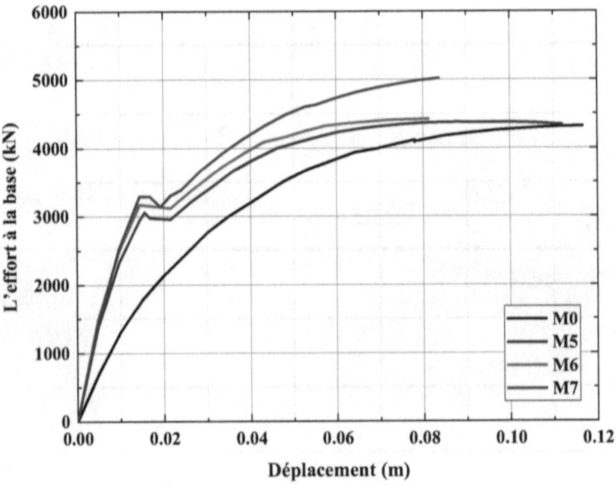

Figure 4.8 Capacity curves

In addition, the percentage of energy absorbed in models M6 and M7 (see Figure 4.9), which were 24% and 16% respectively, was lower than in model M0. However, model M5 recorded an increase of 8%. Therefore, it can be noted that it is important to have infill walls on both sides of a building to reduce the amount of energy absorbed by the building in order to avoid collapse or deterioration. In addition, the presence of masonry infill walls in the various floors of the structure, distributed on both sides of the building, helps to reduce the energy absorbed.

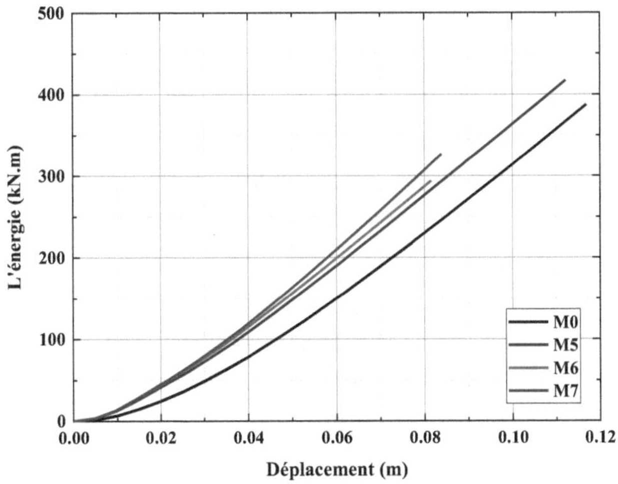

Figure 4.9 Absorbed energy

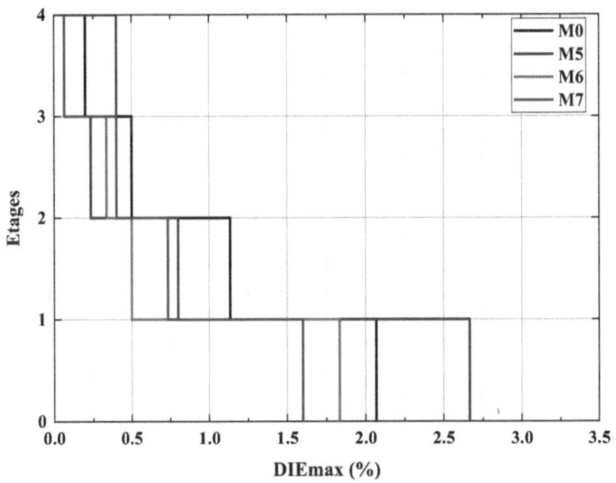

Figure 4.10 Maximum inter-storey displacement

In addition, there is a proportional relationship between the energy absorbed in the building and the DIEmax (see Figure 4.10). There is a 29% increase between models M5 and M0, while models M6 and M7 show a decrease of 11% and 23%, respectively. In addition, there was a 16% increase in the maximum shear force at the base of model M7 compared with model M0, in contrast to the insignificant increase in models M5 and M6. This is due to the absence of masonry infill on the same floors and its random distribution on both sides of the structure.

It can also be seen in Figure 4.11 that the performance level of model M7 is CP. The performance level is LS for models M0, M5 and M6, due to the huge shear force at the base of model M7. Therefore, the distribution and availability of the masonry infill on the different floors of the structure has a beneficial effect in increasing the stiffness and reducing the displacement of the structure.

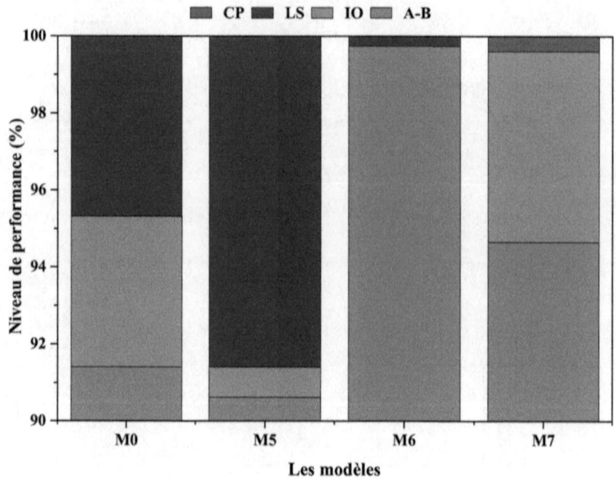

Figure 4.11 Performance levels

In addition, the damage states for these models in the first and last axes, including the arrangement of the infill walls in plan and/or elevation, are presented in Table 4.8.

Table 4.8 Damage states

Models	First axis (Axis 1)	Last axis (Axis 6)
M0	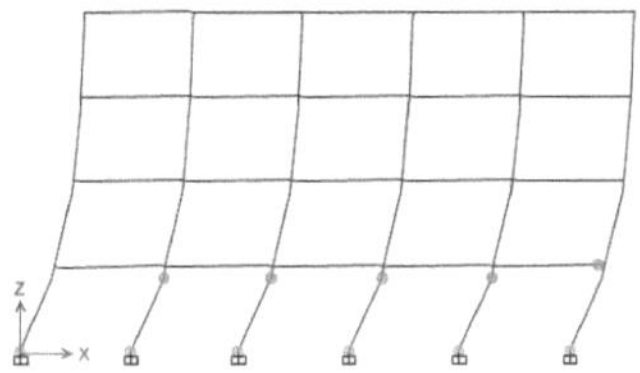	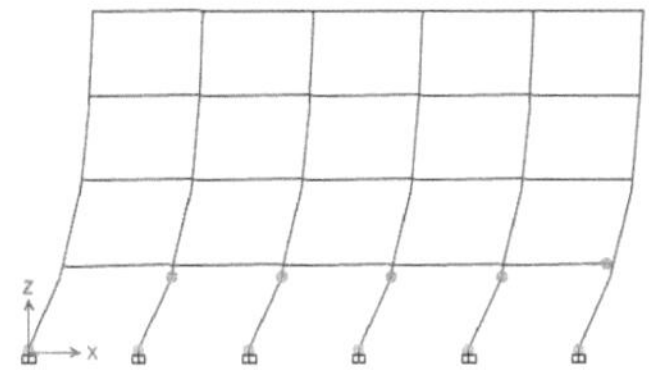
M5	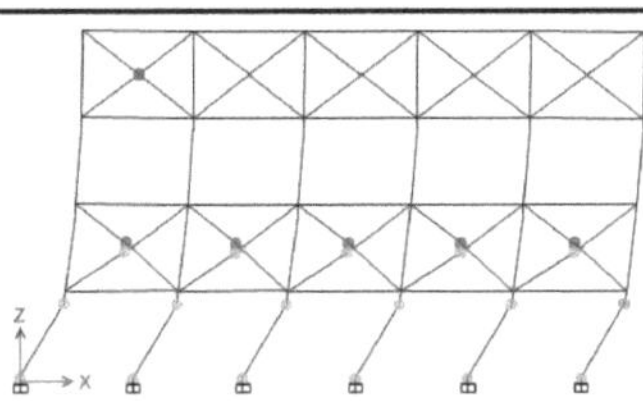	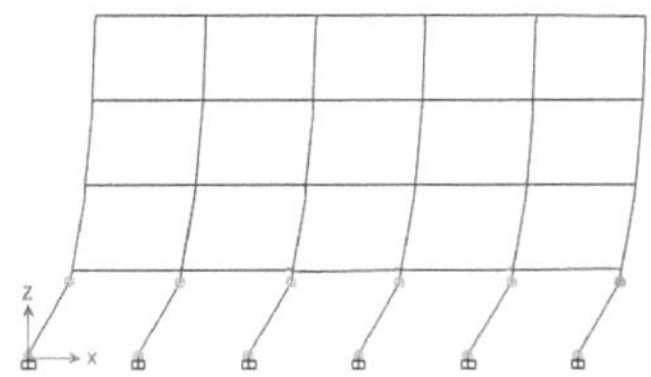
M6	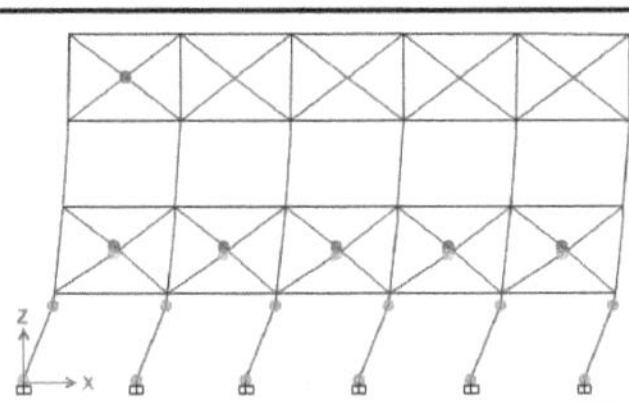	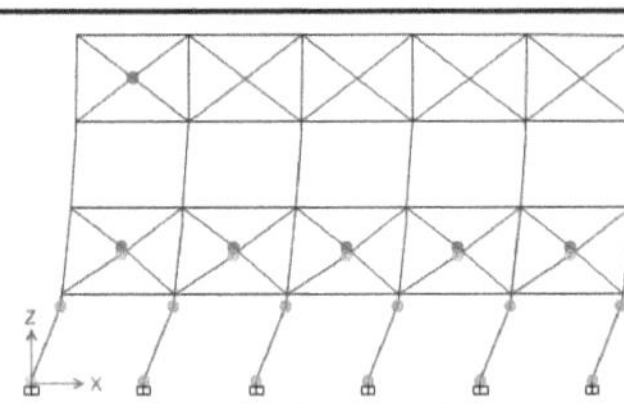
M7	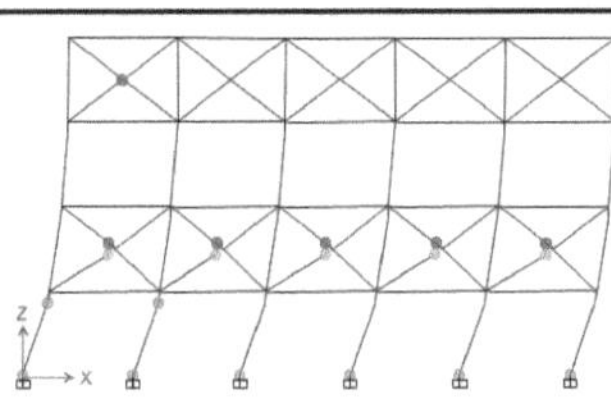	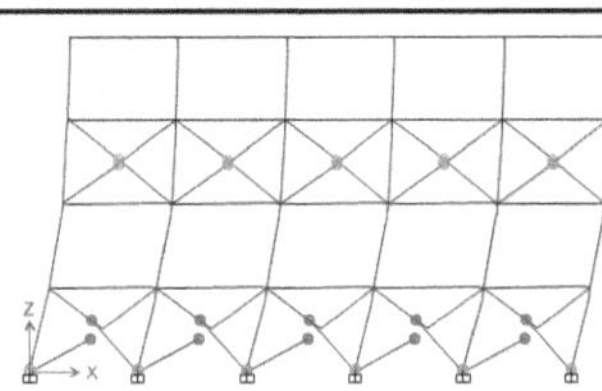

4.5.1.3. Third case study

As shown in Figure 4.12, there is an increase in lateral stiffness for models M8, M9, M10, M11 and M12 compared to model M0 with the following percentages: 111, 121, 208, 149 and 124%, respectively. These increases are due to the presence of infill walls on all floors of the building, with greater increases observed for models M10 and M11, compared to models M8 and M9, respectively, due to the presence of infill walls in consecutive compartments of the central frame, while filling only the side compartments. It can also be seen that there is a reduction in the maximum displacement value of 22, 32 and 31% for models M9, M10 and M12, respectively. On the other hand, models M8 and M10 show a slight increase in maximum displacement of around 5%, compared to model M0, as there are no masonry infill walls on the sides of the structure and in all the compartments.

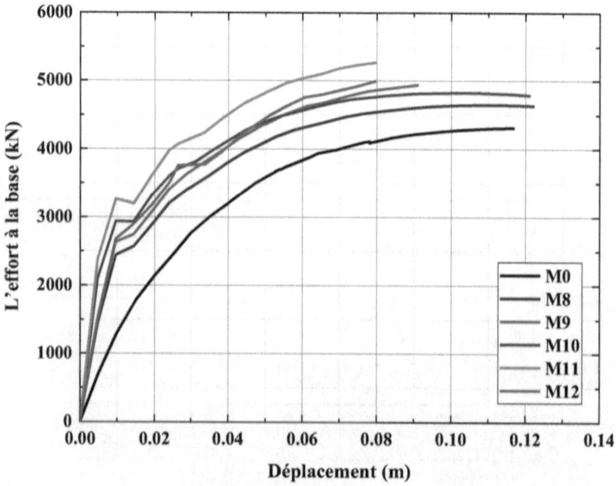

Figure 4.12 Capacity curves

Figure 4.13 shows that models M9, M11 and M12 have absorbed 9, 13 and 19% less energy, respectively, than model M0, while models M8 and M10 have seen increases of 23 and 30%, respectively. This means that it is very important to have infill walls on the sides of the building to reduce the energy absorbed.

This also affects the maximum value of the EIS, which changes in proportion to the absorbed energy of the different models, with a decrease observed for models M9, M11 and M12 of 23, 31 and 35%, respectively, and an increase for models M8 and M10 of 32 and 53%, respectively (see Figure 4.14). In addition, an increase in the maximum shear forces at the base for all models, relative to model M0, is recorded, with the magnitude of the increase depending on the number of infill walls present, ranging from 8 to 22%.

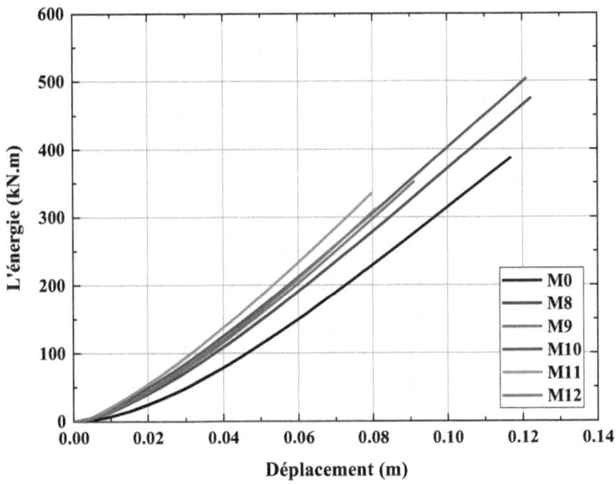

Figure 4.13 Absorbed energy

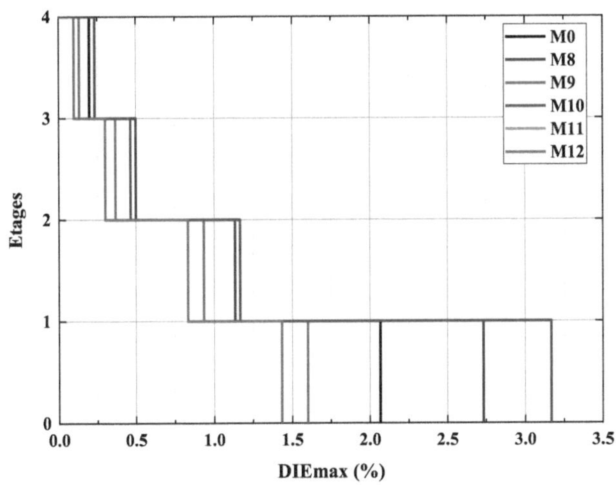

Figure 4.14 Maximum inter-storey displacement

It should also be noted that the performance level of model M0 is LS (see Figure 4.15); it can be seen that the presence of infill walls on the sides of the building leads to a reduction in the level of damage, as can be seen in models M9 and M11, which achieve performance level IO without any change, compared with the other models.

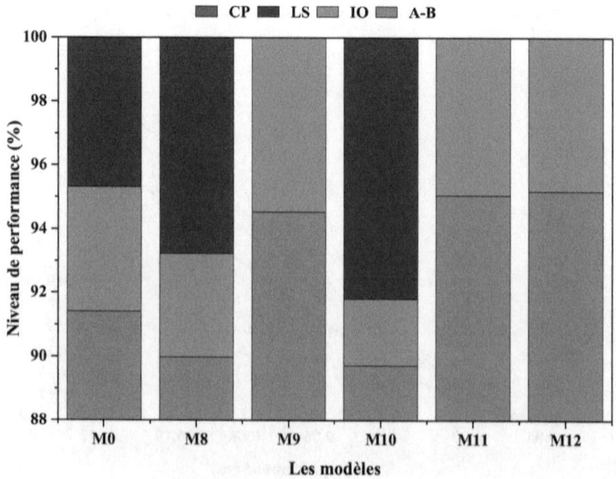

Figure 4.15 Performance levels

In addition, the damage states for these models in the first and last axes, which incorporate the layout of the infill walls in plan and/or elevation, are detailed in Table 4.9.

Table 4.9 Damage states

Models	First axis (Axis 1)	Last axis (Axis 6)
M0	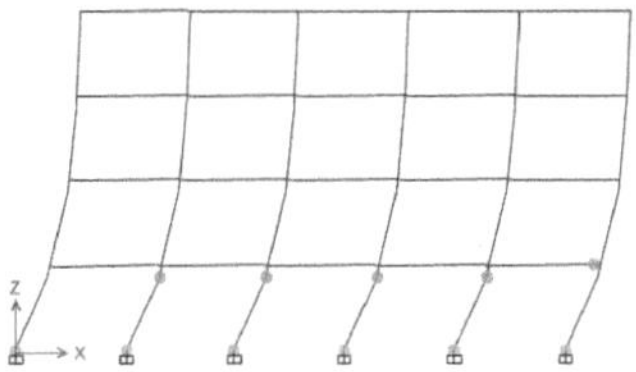	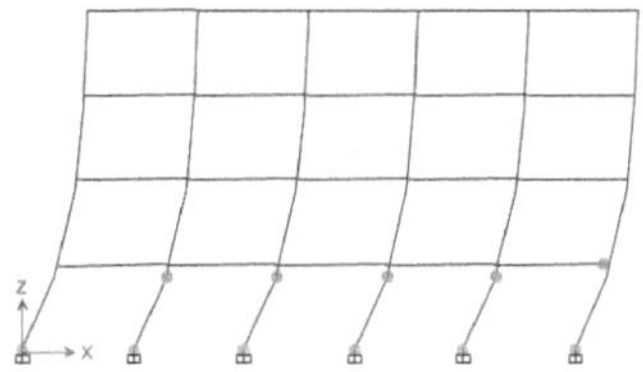
M8	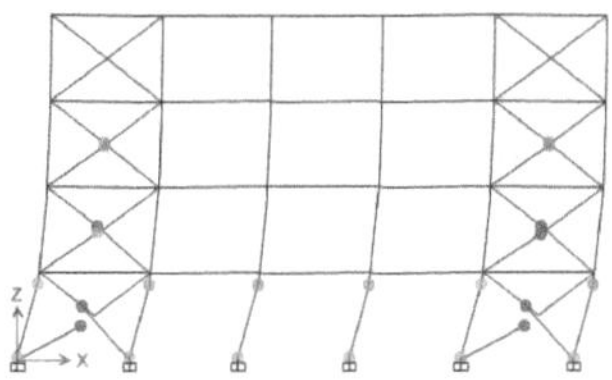	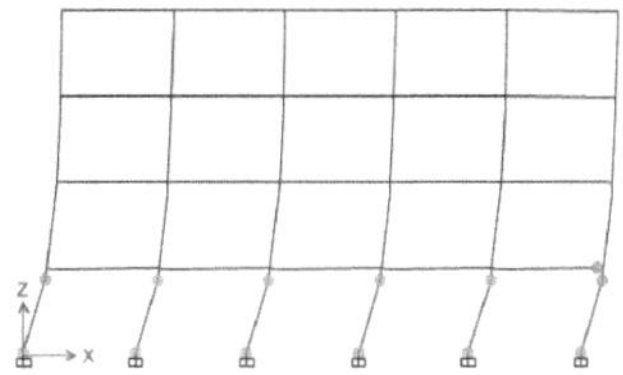
M9	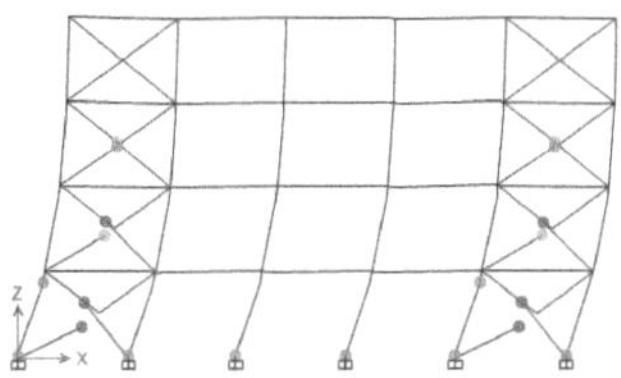	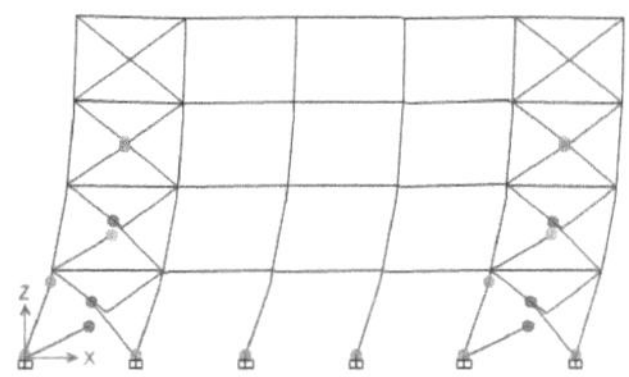
M10	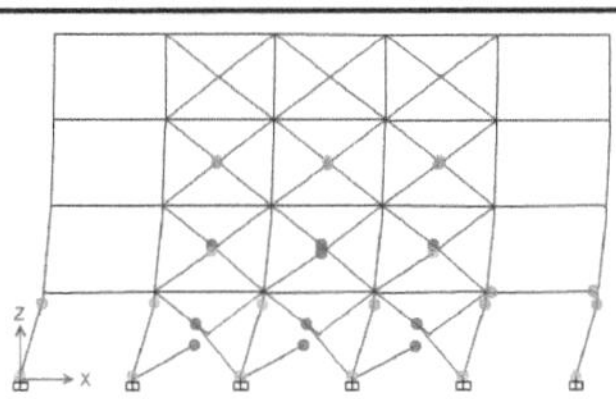	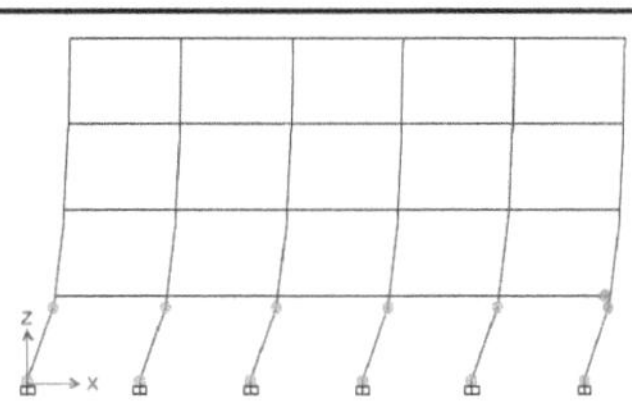
M11	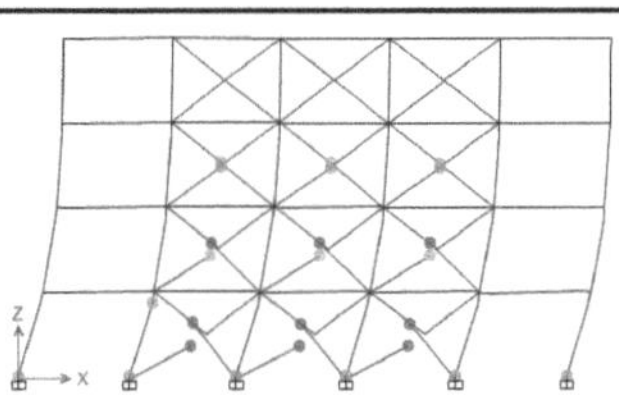	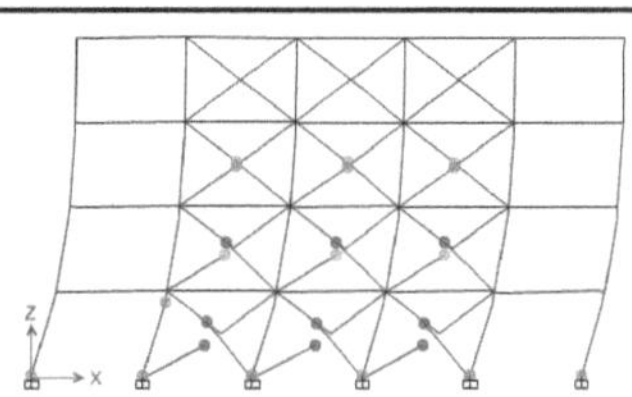
M12	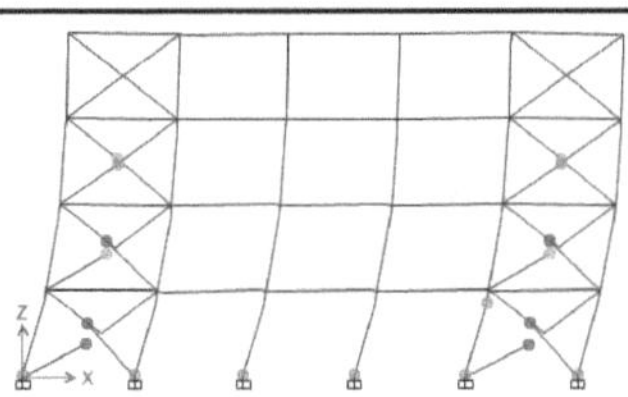	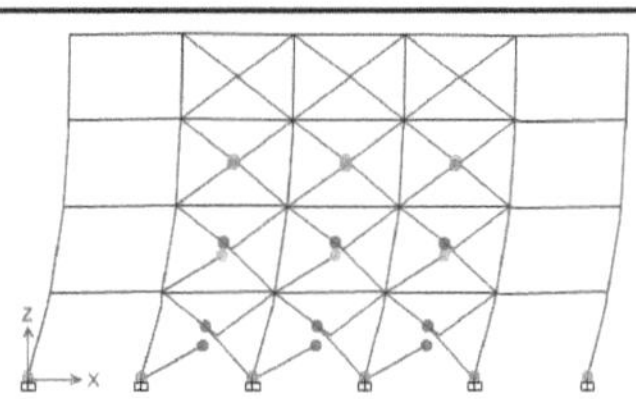

The results of the three cases, using radar diagrams, are presented in Figure 4.16 to compare the results studied, since the values of the results of the models are linked to the values of the results of the building without masonry infill walls; this means that there are no contradictions that prevent us from seeing the resulting differences between all the cases. It can be seen that the value of the maximum displacement of a building is often lower than that of the bare M0 model, with the exception of the slight increase observed for the M8 and M10 models, with the possibility of knowing more about the value of the maximum shear at the base that targets the building. It is also clear that the placement and distribution of the masonry infill walls and their impact give a better reading of the expected performance of the building during an earthquake, based on the readings of the absorbed energy and the DIEmax. It can be seen that all the models with an increase in the absorbed energy and DIEmax values have an asymmetrical layout of the infill walls in plan and/or elevation.

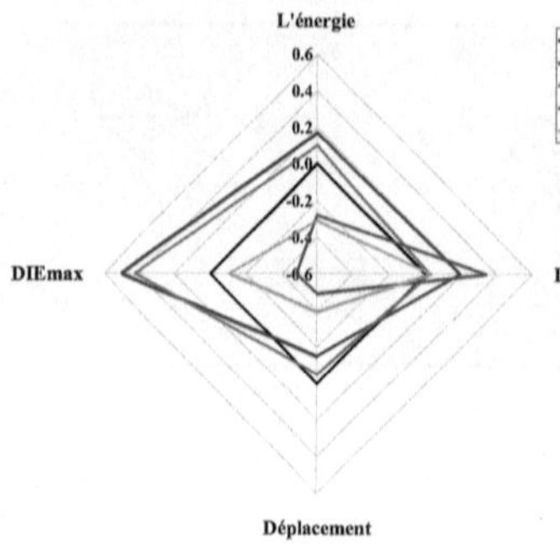

(a) 1st case study

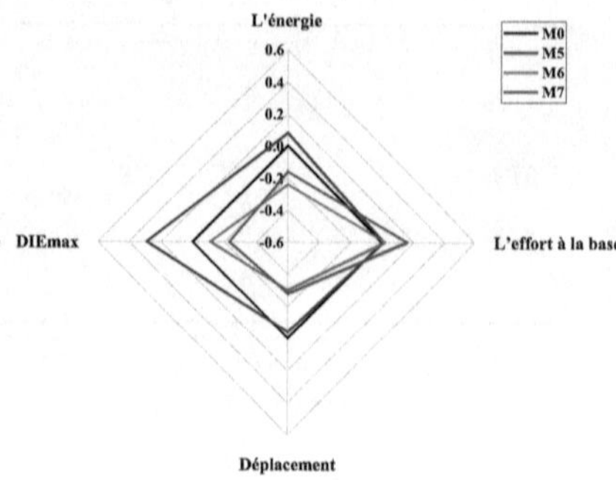

(b) 2nd case study

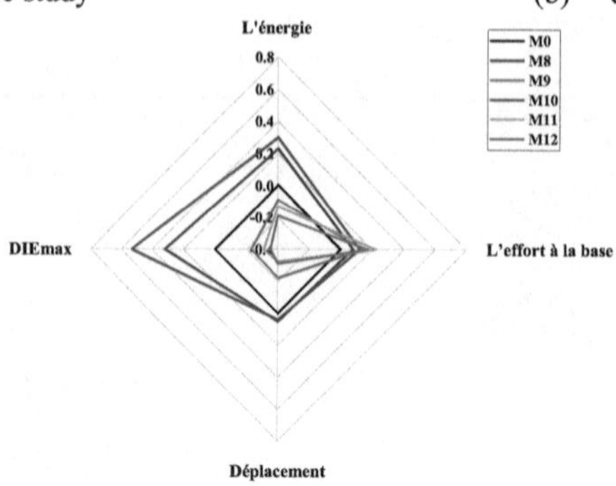

(c) 3rd case study

Figure 4.16 Radar diagrams to compare the results studied in the three cases

4.5.2. Torsion in the structure

In this section, the results presented in Figure 4. 17and obtained from the linear dynamic (spectral) analysis performed on the thirteen modes in the X direction, will be discussed, using the torsion coefficient $C_T$to check whether the structure is subject to torsion or not. In brief, all the models that had symmetry in the distribution of masonry infills on both sides of the building on each floor separately had a torsion coefficient (C_T= 1), meaning that these masonry infills did not add additional torsional force to the building, including model M6, which contains masonry infills symmetrically on the second and fourth floors and their absence on the first and third floors. In contrast to model M7, where the masonry infills are not symmetrically distributed on all floors, it is noted that the first and third floors have a torsion coefficient of more than 120.. Furthermore, with regard to the vertical distribution in model M12, it is noted that there is an additional torsional force ($C_{(T)} <_1$.20), and this is due to the asymmetry in terms of the number of masonry infills on both sides of the building, indicating that the random distribution of masonry infills may result in additional torsional forces for the building that were not taken into account when analysing buildings without masonry infills.

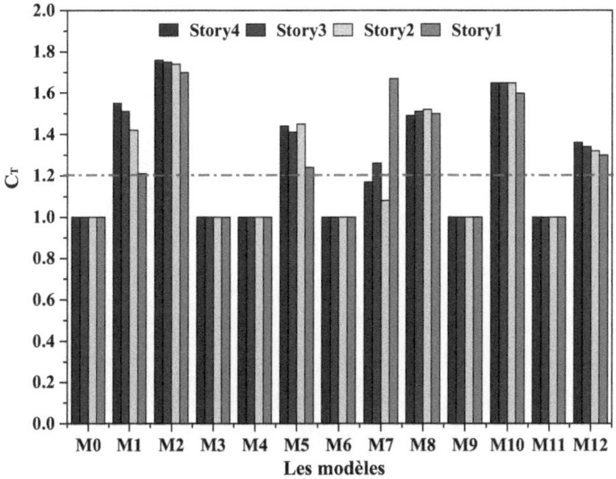

Figure 4. 17Torsion in the structure

4.6. Conclusion

This chapter evaluates the effects of infill wall arrangement in plan and/or elevation on the seismic performance of a 3D reinforced concrete structure located in a high seismicity zone.

A non-linear static analysis procedure (Pushover) is carried out to evaluate the seismic performance of the reinforced concrete structure with different infill wall arrangements, and to study the results related to capacity curves, absorbed energy, inter-storey displacement and performance levels. Next, it is also necessary to check their effect on torsion in the structure. In order to verify whether or not the structure is subject to torsion, a linear dynamic analysis was used. In addition, macro modelling based on fibre sections was used to model the infill walls.

The results show that the presence of infill walls in the building increases lateral stiffness and modifies structural behaviour. The results also show that it is necessary to take into account the effects caused by the layout of the infill walls, as a non-uniform layout of the infill walls can lead to a deterioration in the performance level of the building after an earthquake. In fact, an increase in the energy absorbed leads to an increase in inter-storey displacement and the risk of deterioration or collapse. To avoid this, buildings must have a uniform arrangement of infill walls. In addition, the random arrangement and asymmetry of the number of masonry infill walls on both sides of the building reveals an additional torsional force in the building. The lack of masonry walls on the first floor can lead to irregular rigidity and a "soft floor" phenomenon.

General conclusion

This book undertakes a comprehensive assessment of the effects of the arrangement of infill walls in plan and/or elevation on the seismic performance of reinforced concrete structures. To this end, a methodological approach based on numerical investigation was adopted. Initially, masonry infill walls were modelled using the macro-modelling approach with a discretised fibre section (fibre ball joint) to represent their non-linear behaviour, which enabled a better understanding of their response under horizontal loading when embedded in reinforced concrete frames. Subsequently, non-linear analyses were carried out using ETABS finite element software, with validation by comparison with an experimental study based on force-displacement curves. The aim of this study was to determine the impact of the distribution of infill walls on the seismic behaviour of reinforced concrete structures. The case study method was chosen to facilitate understanding and assimilation of the results. From the conclusions of this research, it is possible to draw several significant observations:

- ✓ The presence of infill walls in structures increases the lateral stiffness of the building, which often leads to a reduction in the maximum displacement when subjected to an earthquake, thus modifying the behaviour of the structure.
- ✓ The lack of masonry walls on the first floor can lead to irregular rigidity and a "soft floor" phenomenon.
- ✓ A uniform arrangement of infill walls in plan and/or elevation plays a beneficial role in improving building performance after an earthquake.
- ✓ The increase in absorbed energy has a negative effect on the maximum inter-storey displacement, generating major risks of the building collapsing or a high rate of deterioration of its components; to avoid this, no side of the building should be left bare of infill walls.
- ✓ The building's performance level improves when the infill walls are present on both sides of the different storeys, independently of the infill of all the bays.
- ✓ Introducing the infill wall modelling process into the analysis steps helps to learn more about the value of the maximum shear force at the base that targets the building.
- ✓ The random arrangement and asymmetry of the number of masonry infill walls on either side of the building reveal an additional twisting force in the building.

In conclusion, this study provides valuable information on the impact of the layout of infill walls on the seismic performance of reinforced concrete structures. These results can be used to guide the design and construction of more seismically resilient buildings, thereby contributing to the safety and sustainability of urban infrastructure. Furthermore, by incorporating the findings of this research into design and construction practices, it is possible to enhance the safety and durability of buildings, thereby ensuring the protection and well-being of occupants as well as the preservation of material assets. Consideration of the implications of this study can also play a crucial role in reducing the economic and social losses caused by structural damage during seismic disasters. By adopting improved construction standards based on the knowledge acquired, it is possible to minimise the risks to human life and ensure the durability of urban infrastructures in the face of seismic challenges.

Finally, applying the recommendations arising from this research can help to strengthen the resilience of urban communities and promote sustainable development in regions at high risk of earthquakes, thereby providing a safer and more secure environment for future generations.

Recommendations for future research

This book has studied the effects of the arrangement of infill walls in plan and/or elevation on the seismic performance of reinforced concrete structures. Nevertheless, some points are not fully covered. Consequently, these points are identified for further investigation:

- Develop a macro model to model the behaviour of infill walls, taking into account openings in the walls.
- Study the effect of the distribution of infill walls in buildings with walls on their performance.
- Carry out studies on buildings with infill walls using SeismoStruct software.

References

[1] Islam MdM, Sen D, Chowdhury SR. Numerical investigation on the effect of infill masonry on lateral behaviour of surrounding RC frame. Asian J Civ Eng 2023. https://doi.org/10.1007/s42107-023-00679-1.

[2] Panthi B, Dahal P, Shrestha P, Thapa KB. Fundamental period of RC buildings with infill walls in Nepal. Asian J Civ Eng 2021;22:983-93. https://doi.org/10.1007/s42107-021-00359-y.

[3] Abdelaziz MM, Gomma MS, El-Ghazaly H. Seismic evaluation of reinforced concrete structures infilled with masonry infill walls. Asian J Civ Eng 2019;20:961-81. https://doi.org/10.1007/s42107-019-00158-6.

[4] Mehrabi AB, Shing PB, Schuller MP, Noland JL. Experimental evaluation of masonry-infilled RC frames. J Struct Eng 1996;122:228-37. https://doi.org/10.1061/(asce)0733-9445(1996)122:3(228).

[5] Mehrabi AB, Shing PB. Finite element modeling of masonry-infilled RC frames. J Struct Eng 1997;123:604-13. https://doi.org/10.1061/(asce)0733-9445(1997)123:5(604).

[6] Shing PB, Mehrabi AB. Behaviour and analysis of masonry-infilled frames. Prog Struct Eng Mater 2002;4:320-31. https://doi.org/10.1002/pse.122.

[7] Al-Chaar G, Issa M, Sweeney S. Behavior of masonry-infilled nonductile reinforced concrete frames. J Struct Eng 2002;128:1055-63. https://doi.org/10.1061/(asce)0733-9445(2002)128:8(1055).

[8] Stylianidis KC. Experimental investigation of masonry infilled R/C frames. Open Constr Build Technol J 2012;6:194-212. https://doi.org/10.2174/1874836801206010194.

[9] Anić F, Penava D, Guljaš I, Sarhosis V, Abrahamczyk L. Out-of-plane cyclic response of masonry infilled RC frames: An experimental study. Eng Struct 2021;238:112258. https://doi.org/10.1016/j.engstruct.2021.112258.

[10] Dolšek M, Fajfar P. The effect of masonry infills on the seismic response of a four-storey reinforced concrete frame - a deterministic assessment. Eng Struct 2008;30:1991-2001. https://doi.org/10.1016/j.engstruct.2008.01.001.

[11] Crisafulli FJ. Seismic behaviour of reinforced concrete structures with masonry infills 1997. http://dx.doi.org/10.26021/1979.

[12] El-Dakhakhni WW, Elgaaly M, Hamid AA. Three-strut model for concrete masonry-infilled steel frames. J Struct Eng 2003;129:177-85. https://doi.org/10.1061/(asce)0733-9445(2003)129:2(177).

[13] Milanesi RR, Morandi P, Magenes G. Local effects on RC frames induced by AAC masonry infills through FEM simulation of in-plane tests. Bull Earthq Eng 2018;16:4053-80. https://doi.org/10.1007/s10518-018-0353-5.

[14] Furtado A, Rodrigues H, Arêde A, Varum H. Simplified macro-model for infill masonry walls considering the out-of-plane behaviour. Earthq Eng Struct Dyn 2015;45:507-24. https://doi.org/10.1002/eqe.2663.

[15] Butenweg C, Marinković M, Salatić R. Experimental results of reinforced concrete frames with masonry infills under combined quasi-static in-plane and out-of-plane seismic loading. Bull Earthq Eng 2019;17:3397-422. https://doi.org/10.1007/s10518-019-00602-7.

[16] Liberatore L, AlShawa O. On the application of the yield-line method to masonry infills subjected to combined in-plane and out-of-plane loads. Structures 2021;32:1287–301. https://doi.org/10.1016/j.istruc.2021.03.044.

[17] Paulay T, Priestley MN. Seismic design of reinforced concrete and masonry buildings. vol. 768. Wiley New York; 1992.

[18] Braga F, Manfredi V, Masi A, Salvatori A, Vona M. Performance of non-structural elements in RC buildings during the L'Aquila, 2009 earthquake. Bull Earthq Eng 2011;9:307-24. https://doi.org/10.1007/s10518-010-9205-7.

[19] Al Hanoun M, Abrahamczyk L, Schwarz J. Numerical simulation of RC frame structures with infill walls under consideration of out-of-plane behavior. Proc. 16th World Conf. Earthq. Eng. WCEE Santiago Chile, 2017, p. 9-13.

[20] Yuen YP, Kuang JS. Fourier-based incremental homogenisation of coupled unilateral damage-plasticity model for masonry structures. Int J Solids Struct 2013;50:3361-74. https://doi.org/10.1016/j.ijsolstr.2013.06.001.

[21] Yuen YP, Kuang JS. Masonry-infilled RC frames subjected to combined in-plane and out-of-plane loading. Int J Struct Stab Dyn 2014;14:1350066. https://doi.org/10.1142/s0219455413500661.

[22] Yuen YP, Kuang JS. Nonlinear seismic responses and lateral force transfer mechanisms of RC frames with different infill configurations. Eng Struct 2015;91:125-40. https://doi.org/10.1016/j.engstruct.2015.02.031.

[23] Yuen TYP, Kuang JS, Ali BSM. Assessing the effect of bi-directional loading on nonlinear static and dynamic behaviour of masonry-infilled frames with openings. Bull Earthq Eng 2016;14:1721-55. https://doi.org/10.1007/s10518-016-9899-2.

[24] Morandi P, Hak S, Magenes G. Simplified out-of-plane resistance verification for slender clay masonry infills in RC frames. Proc XV ANIDIS L'Ingegneria Sismica Ital Padua Italy 2013;30.

[25] Donà M, Tecchio G, Domenicale L, Saler E, Minotto M, Porto F da. Directional effects on combined in-plane and out-of-plane seismic behavior of masonry infills, ECCOMAS Proceedia; 2017. https://doi.org/10.7712/120117.5649.18570.

[26] Asteris PG, Cavaleri L, Trapani FD, Tsaris AK. Numerical modelling of out-of-plane response of infilled frames: State of the art and future challenges for the equivalent strut macromodels. Eng Struct 2017;132:110-22. https://doi.org/10.1016/j.engstruct.2016.10.012.

[27] Di Trapani F, Shing PB, Cavaleri L. Macroelement model for in-plane and out-of-plane responses of masonry infills in frame structures. J Struct Eng 2018;144. https://doi.org/10.1061/(asce)st.1943-541x.0001926.

[28] Polyakov S. Masonry in framed buildings: An investigation into the strength and stiffness of masonry infilling, Translation into English by GL Cairns. Gosdarstvenoeizdatel'stvoliteraturypostroitel'stvuiarkhitekture Mosc 1956.

[29] Sachanski S. Analysis of earthquake resistance of frame buildings taking into consideration the carring capacity of the filling masonry. Proc. 2nd World Conf. Earthq. Eng. vol. 138, 1960, p. 1-15.

[30] Holmes M. Steel frames with brickwork and concrete infilling. Proc Inst Civ Eng 1961;19:473-8. https://doi.org/10.1680/iicep.1961.11305.

[31] Stafford Smith B. Lateral stiffness of infilled frames. J Struct Div 1962;88:183-99. https://doi.org/10.1061/JSDEAG.0000849.

[32] Fiorato AE, Sozen MA, Gamble WL. An Investigation of the Interaction of Reinforced Concrete Frames with Masonry Filler Walls. Civ Eng Stud SRS- 370 1970.

[33] Mainstone RJ. On the stiffness and strengths of infilled frames. Proc Inst Civ Eng 1971;49:230. https://doi.org/10.1680/iicep.1971.6267.

[34] Mainstone RJ, Weeks G. The influence of a bounding frame on the racking stiffness and strengths of brick walls. Building Research Station; 1972.

[35] Klingner RE, Bertero VV. Earthquake resistance of infilled frames. J Struct Div 1978;104:973-89. https://doi.org/10.1061/jsdeag.0004947.

[36] Zarnic R, Tomazevic M. An experimentally obtained method for evaluation of the behavior of masonry infilled RC frames. Proc. 9th World Conf. Earthq. Eng, vol. 6, 1988, p. 163-8.

[37] Pires F. Influence of masonry walls over the behaviour of reinforced concrete frames under horizontal actions. PhD Thesis. National Laboratory in Civil Engineering Lisbon, Portugal, 1990.

[38] Mehrabi AB. Behavior of masonry infilled reinforced concrete frames subjected to lateral loadings. 1994.

[39] Mosalam KM, White RN, Ayala G. Response of infilled frames using pseudo-dynamic experimentation. Earthq Eng Struct Dyn 1998;27:589-608.

[40] Buonopane S, White R. Pseudodynamic testing of masonry infilled reinforced concrete frame. J Struct Eng 1999;125:578-89.

[41] Lee H-S, Woo S-W. Effect of masonry infills on seismic performance of a 3-storey R/C frame with non-seismic detailing. Earthq Eng Struct Dyn 2002;31:353-78.

[42] Süsoy M. Seismic strengthening of masonry infilled reinforced concrete frames with precast concrete panels. Master's Thesis. Middle East Technical University, 2004.

[43] Hashemi A, Mosalam KM. Shake-table experiment on reinforced concrete structure containing masonry infill wall. Earthq Eng Struct Dyn 2006;35:1827-52.

[44] Pinto A, Taucer F. Assessment and retrofit of full-scale models of existing RC frames. Adv. Earthq. Eng. Urban Risk Reduct, Springer; 2006, p. 353-67.

[45] Anil Ö, Altin S. An experimental study on reinforced concrete partially infilled frames. Eng Struct 2007;29:449-60.

[46] Kakaletsis DJ, Karayannis CG. Influence of masonry strength and openings on infilled R/C frames under cycling loading. J Earthq Eng 2008;12:197-221.

[47] Kakaletsis DJ, Karayannis CG. Experimental investigation of infilled reinforced concrete frames with openings. Struct J 2009;106:132-41. https://doi.org/10.14359/56351.

[48] Blackard B, Willam K, Mettupalayam S. Experimental observations of masonry infilled reinforced concrete frames with openings. Spec Publ 2009;265:199-122.

[49] Stavridis A, Koutromanos I, Shing PB. Shake-table tests of a three-story reinforced concrete frame with masonry infill walls. Earthq Eng Struct Dyn 2012;41:1089-108.

[50] Misir S, Ozcelik O, Girgin SC, Kahraman S. Experimental work on seismic behavior of various types of masonry infilled RC frames. Struct Eng Mech 2012;44:763-74.

[51] Sigmund V, Penava D. Assessment of masonry infilled reinforced-concrete frames with openings. Teh Vjesn 2013;20:459-66.

[52] Mansouri A, Marefat MS, Khanmohammadi M. Experimental evaluation of seismic performance of low-shear strength masonry infills with openings in reinforced concrete frames with deficient seismic details. Struct Des Tall Spec Build 2014;23:1190-210. https://doi.org/10.1002/tal.1115.

[53] Bergami AV, Nuti C. Experimental tests and global modeling of masonry infilled frames. Earthq Struct 2015;9:281-303.

[54] Basha SH, Kaushik HB. Behavior and failure mechanisms of masonry-infilled RC frames (in low-rise buildings) subject to lateral loading. Eng Struct 2016;111:233-45.

[55] Shan S, Li S, Xu S, Xie L. Experimental study on the progressive collapse performance of RC frames with infill walls. Eng Struct 2016;111:80-92.

[56] Zhai C, Kong J, Wang X, Chen Z. Experimental and finite element analytical investigation of seismic behavior of full-scale masonry infilled RC frames. J Earthq Eng 2016;20:1171-98. https://doi.org/10.1080/13632469.2016.1138171.

[57] Oliveira DV, Lourenço PB. Implementation and validation of a constitutive model for the cyclic behaviour of interface elements. Comput Struct 2004;82:1451-61.

[58] Crisafulli FJ, Carr AJ, Park R. Analytical modelling of infilled frame structures: A general review. Bull N Z Soc Earthq Eng 2000;33:30-47. https://doi.org/10.5459/bnzsee.33.1.30-47.

[59] Asteris PG, Antoniou ST, Sophianopoulos DS, Chrysostomou CZ. Mathematical macromodeling of infilled frames: state of the art. J Struct Eng 2011;137:1508-17. https://doi.org/10.1061/(asce)st.1943-541x.0000384.

[60] Nicola T, Leandro C, Guido C, Enrico S. Masonry infilled frame structures: state-of-the-art review of numerical modelling. Earthq Struct 2015;8:733-59. https://doi.org/10.12989/eas.2015.8.3.733.

[61] Petracca M, Pelà L, Rossi R, Oller S, Camata G, Spacone E. Multiscale computational first order homogenization of thick shells for the analysis of out-of-plane loaded masonry walls. Comput Methods Appl Mech Eng 2017;315:273-301. https://doi.org/10.1016/j.cma.2016.10.046.

[62] Riddington JR, Naom NF. Finite element prediction of masonry compressive strength. Comput Struct 1994;52:113-9. https://doi.org/10.1016/0045-7949(94)90261-5.

[63] Lourenço PB, Rots JG. Multisurface interface model for analysis of masonry structures. J Eng Mech 1997;123:660-8. https://doi.org/10.1061/(asce)0733-9399(1997)123:7(660).

[64] Al-Chaar G, Mehrabi AB, Manzouri T. Finite element interface modeling and experimental verification of masonry-infilled R/C frames. Masonry Society Longmont, CO, USA; 2008.

[65] Stavridis A, Shing P. Finite-element modeling of nonlinear behavior of masonry-infilled RC frames. J Struct Eng 2010;136:285-96.

[66] Koutromanos I, Stavridis A, Shing PB, Willam K. Numerical modeling of masonry-infilled RC frames subjected to seismic loads. Comput Struct 2011;89:1026-37.

[67] Stavridis A. Analytical and experimental study of seismic performance of reinforced concrete frames infilled with masonry walls. University of California, San Diego; 2009.

[68] Sattar S. Influence of masonry infill walls and other building characteristics on seismic collapse of concrete frame buildings. PhD Thesis. University of Colorado at Boulder, 2013.

[69] Mohyeddin A, Goldsworthy HM, Gad EF. FE modelling of RC frames with masonry infill panels under in-plane and out-of-plane loading. Eng Struct 2013;51:73-87.

[70] Al Louzi RAK. Seismic in-plane response of reinforced concrete frames with masonry infill walls. PhD Thesis. Purdue University, 2015.

[71] Kubalski T, Marinković M, Butenweg C. Numerical investigation of masonry infilled RC frames. 16th Int. Brick Block Mason. Conf. 2016, pp. 26-30.

[72] Quinteros RD, Oller S, Nallim LG. Nonlinear homogenization techniques to solve masonry structures problems. Compos Struct 2012;94:724-30. https://doi.org/10.1016/j.compstruct.2011.09.006.

[73] Lopez J, Oller S, Oñate E, Lubliner J. A homogeneous constitutive model for masonry. Int J Numer Methods Eng 1999;46:1651-71. https://doi.org/10.1002/(sici)1097-0207(19991210)46:10<1651::aid-nme718>3.0.co;2-2.

[74] Ma G, Hao H, Lu Y. Homogenization of masonry using numerical simulations. J Eng Mech 2001;127:421-31. https://doi.org/10.1061/(asce)0733-9399(2001)127:5(421).

[75] Milani G, Lourenço PB, Tralli A. Homogenised limit analysis of masonry walls, Part I: Failure surfaces. Comput Struct 2006;84:166-80. https://doi.org/10.1016/j.compstruc.2005.09.005.

[76] Milani G, Lourenço PB, Tralli A. Homogenised limit analysis of masonry walls, Part II: Structural examples. Comput Struct 2006;84:181-95. https://doi.org/10.1016/j.compstruc.2005.09.004.

[77] Milani G, Lourenço PB. FE homogenised limit analysis model for masonry structures. Proc Inst Civ Eng - Eng Comput Mech 2011;164:65-78. https://doi.org/10.1680/eacm.2011.164.2.65.

[78] Komurcu S, Gedikli A. Numerical modelling of the in-plane loaded homogenized masonry walls. Period Eng Nat Sci PEN 2017;5. https://doi.org/10.21533/pen.v5i3.117.

[79] Tiberti S, Milani G. 3D homogenized limit analysis of non-periodic multi-leaf masonry walls. Comput Struct 2020;234:106253. https://doi.org/10.1016/j.compstruc.2020.106253.

[80] Tiberti S, Milani G. 3D voxel homogenized limit analysis of single-leaf non-periodic masonry. Comput Struct 2020;229:106186. https://doi.org/10.1016/j.compstruc.2019.106186.

[81] Yu S, Wu C, Griffith M. Numerical analysis of out-of-plane loaded masonry wall using homogenization technique. 14th Int. Brick Block Mason. Conf. Syd., 2008, p. 17-20.

[82] Sacco E. A nonlinear homogenization procedure for periodic masonry. Eur J Mech-ASolids 2009;28:209-22.

[83] Houda F, Myriam HL, Ioannis S, Sab K, Dogui A. In-Plane Strength Domain Numerical Determination of Hollow Concrete Block Masonry. Des. Model. Mech. Syst. Proc. 7th

Conf. Des. Model. Mech. Syst. CMSM2017 March 27-29 Hammamet Tunis. 7, Springer; 2018, pp. 721-32.

[84] Polyakov S. On the interaction between masonry filler walls and enclosing frame when loaded in the plane of the wall. Transl Earthq Eng 1960;2:36-42.

[85] Stafford Smith B. Behavior of square infilled frames. J Struct Div 1966;92:381-404. https://doi.org/10.1061/jsdeag.0001387.

[86] FEMA 356. Prestandard and commentary for seismic rehabilitation of buildings. Fed Emerg Manag Agency Washigton DC Md USA 2000.

[87] Mucedero G, Perrone D, Brunesi E, Monteiro R. Numerical modelling and validation of the response of masonry infilled rc frames using experimental testing results. Buildings 2020;10:182. https://doi.org/10.3390/buildings10100182.

[88] Leuchars J, Scrivener J. Masonry infill panels subjected to cyclic in-plane loading. Bull N Z Soc Earthq Eng 1976;9:122-31.

[89] Chrysostomou CZ, Gergely P, Abel JF. A six-strut model for nonlinear dynamic analysis of steel infilled frames. Int J Struct Stab Dyn 2002;02:335-53. https://doi.org/10.1142/s0219455402000567.

[90] Mohamed HMA. Seismic risk assessment of reinforced concrete frames with masonry infill. PhD Thesis. Universidade do Porto (Portugal), 2017.

[91] Tucker CJ. Predicting the in-plane capacity of masonry infilled frames. PhD Thesis. Tennessee Technological University, 2007.

[92] Stafford Smith B, Carter C. A METHOD OF ANALYSIS FOR INFILLED FRAMES. Proc Inst Civ Eng 1969;44:31-48. https://doi.org/10.1680/iicep.1969.7290.

[93] Mainstone RJ. Supplementary note on the stiffnesses and strengths of infilled frames. Building Research Establishment, Building Research Station; 1974.

[94] Tassios T. Masonry infill and RC walls,(An invited state-of-the-art report). Third Int. Symp. Wall Struct. 1984.

[95] Bazan E, Meli R. Seismic analysis of structures with masonry walls. Proc 7th World Conf Earthq. Eng, vol. 5, International Association of Earthquake Engineering (IAEE) Tokyo; 1980, p. 633-40.

[96] Te-Chang L, Kwok-Hung K. Nonlinear behaviour of non-integral infilled frames. Comput Struct 1984;18:551-60. https://doi.org/10.1016/0045-7949(84)90070-1.

[97] Decanini L, Fantin G. Modelos simplificados de la mampostería incluida en porticos. Caracteristicas Stiffnessy Resist Lateral En Estado Lte Jorn Argent Ing Estructural 1986;2:817-36.

[98] Moghaddam H. Earthquake resistant design of brick infilled frames. Brick Block Mason 8 Th IBMAC Lond Elsevier Appl Sci 1988;2:774-84.

[99] Durrani AJ, Luo Y. Seismic retrofit of flat-slab buildings with masonry infills. Proc. NCEER Workshop Seism. Response Mason. Infills, 1994, p. 1-8.

[100] Flanagan RD, Bennett RM. Arching of masonry infilled frames: comparison of analytical methods. Pract Period Struct Des Constr 1999;4:105-10. https://doi.org/10.1061/(asce)1084-0680(1999)4:3(105).

[101] Bennett RM, Fischer WL, Flanagan RD, Tenbus MA. Evaluation and analysis of the performance of masonary infills during the Northridge earthquake. Office of Scientific and Technical Information (OSTI); 1996. https://doi.org/10.2172/414635.

[102] Al-Chaar G. Evaluating strength and stiffness of unreinforced masonry infill structures. Defense Technical Information Center; 2002. https://doi.org/10.21236/ada407072.

[103] Chethan K, Babu RR, Venkataramanna K, Sharma A. Study on dynamic characteristics of 3d reinforced concrete frame with masonry infill. Power Res- J CPRI 2009:11-8.

[104] Khaja M. Earthquake analysis on 2d rc frames with different aspect ratios of masonry infill and monolithic panel. Int J Res Eng Technol 2013;02:18-23. https://doi.org/10.15623/ijret.2013.0213004.

[105] Turgay T, Durmus MC, Binici B, Ozcebe G. Evaluation of the predictive models for stiffness, strength, and deformation capacity of RC frames with masonry infill walls. J Struct Eng 2014;140. https://doi.org/10.1061/(asce)st.1943-541x.0001069.

[106] Andreaus U, Cerone M, D'Asdia P, Iannozzi F. A finite element model for the analysis of masonry structures under cyclic actions. Proceeding Seventh Int Brick Mason Conf 1985:479-88.

[107] Doudoumis IN, Mitsopoulou EN. Non-linear analysis of multi-story infilled frames for unilateral contact condition. Earthq Eng 1986:63-70.

[108] Madan A, Reinhorn AM, Mander JB, Valles RE. Modeling of masonry infill panels for structural analysis. J Struct Eng 1997;123:1295-302. https://doi.org/10.1061/(asce)0733-9445(1997)123:10(1295).

[109] Bertoldi S, Decanini L, Gavarini C. Telai tamponati soggetti ad azioni sismiche, un modello semplificato: Confronto sperimentale e numerico. Atti Del 1993;6:815-24.

[110] Panagiotakos TB, Fardis MN. Seismic response of infilled RC frames structures. 11th World Conf. Earthq. Eng. 1996.

[111] Uva G, Raffaele D, Porco F, Fiore A. On the role of equivalent strut models in the seismic assessment of infilled RC buildings. Eng Struct 2012;42:83-94. https://doi.org/10.1016/j.engstruct.2012.04.005.

[112] De Risi MT, Gaudio CD, Ricci P, Verderame GM. In-plane behaviour and damage assessment of masonry infills with hollow clay bricks in RC frames. Eng Struct 2018;168:257-75. https://doi.org/10.1016/j.engstruct.2018.04.065.

[113] Van TC, Lau TL, Mohamed Nazri F. Macro-modeling approach incorporating fiber plastic hinge for reinforced concrete frames with masonry infill. Eng Struct 2022;251:113421. https://doi.org/10.1016/j.engstruct.2021.113421.

[114] Ngo-Huu C, Kim S-E, Oh J-R. Nonlinear analysis of space steel frames using fiber plastic hinge concept. Eng Struct 2007;29:649-57. https://doi.org/10.1016/j.engstruct.2006.06.008.

[115] Ngo-Huu C, Kim S-E. Practical advanced analysis of space steel frames using fiber hinge method. Thin-Walled Struct 2009;47:421-30. https://doi.org/10.1016/j.tws.2008.08.007.

[116] Ngo-huu C, Kim S-E. Practical nonlinear analysis of steel-concrete composite frames using fiber-hinge method. J Constr Steel Res 2012;74:90-7. https://doi.org/10.1016/j.jcsr.2012.02.018.

[117] Wang Y, Wang Q, Dong J, Peng Y. Ultimate analysis of steel structures based on fiber hinge model and time-varying structure theory. Int J Steel Struct 2015;15:567-79.

[118] Wang Y, Peng Y, Dong J. Fiber hinge approach for nonlinear analysis of structural members accounting for local buckling under large deformation. IOP Conf Ser Mater Sci Eng 2018;397:012040. https://doi.org/10.1088/1757-899X/397/1/012040.

[119] Di Trapani F, Bertagnoli G, Ferrotto MF, Gino D. Empirical equations for the direct definition of stress-strain laws for fiber-section-based macromodeling of infilled frames. J Eng Mech 2018;144:04018101. https://doi.org/10.1061/(ASCE)EM.1943-7889.0001532.

[120] Adibi M, Talebkhah R, Ghatte HF. Seismic reliability of precast concrete frame with masonry infill wall. Earthq. Struct. vol. 24, 2023, p. 141-53. https://doi.org/10.12989/eas.2023.24.2.141.

[121] Applied Technology Council. Guidelines for nonlinear structural analysis and design of buildings. part IIb - reinforced concrete moment frames. Gaithersburg, MD: National Institute of Standards and Technology; 2017.

[122] Gupta B. Enhanced pushover procedure and inelastic demand estimation for performance-based seismic evaluation of buildings. University of Central Florida; 1998.

[123] Algériennes RP. Centre National de Recherche Appliquée en Génie Parasismique. Algiers Algér 2003.

[124] Gupta AK. Response spectrum method in seismic analysis and design of structures. Routledge; 2017.

[125] Lee U, Kim J, Leung AYT. The Spectral Element Method in Structural Dynamics. Shock Vib Dig 2000;32:451-65. https://doi.org/10.1177/058310240003200601.

[126] Rooshenas A. Investigating the effects of masonry infill panels on high-rise structures. Structures 2022;35:106–17. https://doi.org/10.1016/j.istruc.2021.10.077.

[127] Adnane O. Nonlinear statistical and dynamic analysis of self-stable three-dimensional structures. PhD Thesis. Béjaia, Université Abderrahmane Mira. Faculty of Sciences and Engineering, 2010.

[128] Belkheir MA. Influence of wall arrangement on the seismic performance of composite structures. PhD Thesis. Université Mouloud Mammeri, 2012.

[129] MENASRI Y, NOUAOURIA M, GUENFOUD M. EVALUATION OF SEISMIC VULNERABILITY AT THE SCALE OF A REINFORCED CONCRETE STRUCTURE USING NON-LINEAR STATIC ANALYSIS (PUSHOVER). Symp. Int. Sur Constr. En Zone Sismique Univ. Guelma Algerie, 2010.

[130] Poursha M, Khoshnoudian F, Moghadam AS. Assessment of modal pushover analysis and conventional nonlinear static procedure with load distributions of federal emergency management agency for high-rise buildings. Struct Des Tall Spec Build 2008:n/a-n/a. https://doi.org/10.1002/tal.487.

[131] Ima M, Benjamin L, Irma J, Hartanto W. Performance Of Modal Pushover Analysis On A First Mode Dominant Moment Resisting Frame. EACEF-Int. Conf. Civ. Eng. vol. 1, 2011, p. 128-128.

[132] Applied Technology Council (ATC-40). Seismic evaluation and retrofit of concrete buildings. ATC-40 1996;1 and 2.

[133] Code P. Eurocode 8: Design of structures for earthquake resistance-part 1: general rules, seismic actions and rules for buildings. Bruss Eur Comm Stand 2005.

[134] Mouroux P, Negulescu C. Practical comparison between the displacement methods of ATC 40 (damping) and Eurocode 8 (ductility). 7th Colloq. Natl. AFPS, 2007.

[135] Elnashai AS. Advanced inelastic static (pushover) analysis for earthquake applications. Struct Eng Mech 2001;12:51-69. https://doi.org/10.12989/sem.2001.12.1.051.

[136] Fajfar P. A Nonlinear Analysis Method for Performance-Based Seismic Design. Earthq Spectra 2000;16:573-92. https://doi.org/10.1193/1.1586128.

[137] Krawinkler H, Seneviratna GDPK. Pros and cons of a pushover analysis of seismic performance evaluation. Eng Struct 1998;20:452-64. https://doi.org/10.1016/s0141-0296(97)00092-8.

[138] Mwafy AM, Elnashai AS. Static pushover versus dynamic collapse analysis of RC buildings. Eng Struct 2001;23:407-24. https://doi.org/10.1016/S0141-0296(00)00068-7.

[139] Paillere A. Metal fibre concrete: current state of knowledge. Ann. Inst. Tech. Bâtim. Trav. Publics, 1993, p. 37-68.

[140] Fajfar P. Capacity spectrum method based on inelastic demand spectra. Earthq Eng Struct Dyn 1999;28:979-93. https://doi.org/10.1002/(SICI)1096-9845(199909)28:9<979::AID-EQE850>3.0.CO;2-1.

[141] Dolšek M, Fajfar P. Simplified non-linear seismic analysis of infilled reinforced concrete frames. Earthq Eng Struct Dyn 2005;34:49-66. https://doi.org/10.1002/eqe.411.

[142] Urbanski A, Spacone E, Belgasmia M, Sarf J, Zimmermann T. Static Pushover Analysis. ZSoil PC n.d.

[143] Freeman SA. Evaluations of existing buildings for seismic risk-A case study of Puget Sound Naval Shipyard. Proc 1st US Nat Conf Earthq. Engrg Bremerton Wash. 1975, 1975, p. 113-22.

[144] Miranda E. Estimation of inelastic deformation demands of SDOF systems. J Struct Eng 2001;127:1005-12.

[145] FEMA 273. NEHRP guidelines for the seismic rehabilitation of buildings. Fed Emerg Manag Agency Wash DC 1997.

[146] FEMA 440, Improvement of nonlinear static seismic analysis procedures. Appl Technol Counc Wash DC 2005.

[147] ASCE/SEI 41-06. Seismic Rehabilitation of Existing Buildings (ASCE/SEI 41-06). Am Soc Civ Eng Rest 2007.

[148] ASCE/SEI 41-13. Seismic Rehabilitation of Existing Buildings (ASCE/SEI 41-13). Am Soc Civ Eng Rest 2013.

[149] Cavaleri L, Di Trapani F. Cyclic response of masonry infilled RC frames: Experimental results and simplified modeling. Soil Dyn Earthq Eng 2014;65:224-42.

[150] Mander JB, Priestley MJ, Park R. Theoretical stress-strain model for confined concrete. J Struct Eng 1988;114:1804-26.

[151] Richart FE, Brandtzæg A, Brown RL. A study of the failure of concrete under combined compressive stresses. Univ Ill Eng Exp Stn Bull No 185 1928.

[152] Popovics S. A numerical approach to the complete stress-strain curve of concrete. Cem Concr Res 1973;3:583-99.

[153] Park R, Paulay T. Reinforced concrete structures. John Wiley & Sons; 1991.

[154] Asteris PG, Repapis CC, Tsaris AK, Di Trapani F, Cavaleri L. Parameters affecting the fundamental period of infilled RC frame structures. Earthq Struct 2015;9:999-1028. https://doi.org/10.12989/EAS.2015.9.5.999.

[155] Papia M, Cavaleri L, Fossetti M. Infilled frames: developments in the evaluation of the stiffening effect of infills. Struct Eng Mech 2003;16:675-93. https://doi.org/10.12989/SEM.2003.16.6.675.

[156] RPA2003. Algerian seismic rules RPA 99/version 2003. Cent Natl Rech Appliquée En Genie Parasismique 2003.

[157] BAEL 91. Règles techniques de conception et de calcul des ouvrages et constructions en béton armé, Suivant la Méthode des états Limites. Cent Sci Tech Bâtim Eyrolles Paris 2001.

[158] Chopra AK, Goel RK. Capacity-demand-diagram methods for estimating seismic deformation of inelastic structures: SDF systems. Rep No PEER199902 1999.

[159] Fajfar P, Gašperšič P. THE N2 METHOD FOR THE SEISMIC DAMAGE ANALYSIS OF RC BUILDINGS. Earthq Eng Struct Dyn 1996;25:31-46. https://doi.org/10.1002/(SICI)1096-9845(199601)25:1<31::AID-EQE534>3.0.CO;2-V.

[160] Rosenblueth E, Meli R. The 1985 mexico earthquake. Concr Int 1986;8:23-34.

[161] Fajfar P. The story of the N2 method. 2022.

[162] American Society of Civil Engineers, editor. Minimum design loads for buildings and other structures. Reston, Va: American Society of Civil Engineers : Structural Engineering Institute; 2010.

FSC
www.fsc.org
MIX
Papier | Fördert
gute Waldnutzung
FSC® C083411

Kolibri 360 GmbH
Ferdinand-Jühlke-Straße 7
99095 Erfurt, Deutschland
produktsicherheit@kolibri360.de

Druck:
CPI Druckdienstleistungen GmbH
im Auftrag der
Kolibri 360 GmbH
Ferdinand-Jühlke-Str. 7
99095 Erfurt